The Other Australia/Japan Living Marine Resources Dispute

The Other Australia/Japan Living Marine Resources Dispute

Inferences on the Merits of the Southern Bluefin Tuna Arbitration in Light of the Whaling Case

By

Andrew Serdy

BRILL

LEIDEN | BOSTON

Library of Congress Control Number: 2016956955

Typeface for the Latin, Greek, and Cyrillic scripts: "Brill". See and download: brill.com/brill-typeface.

ISBN 978-90-04-33944-6 (paperback)
ISBN 978-90-04-33945-3 (e-book)

Originally published as Volume 1(1) 2017, in *The Law of the Sea*, DOI 10.1163/24519359-12340001.

This book is printed on acid-free paper and produced in a sustainable manner.

Contents

The Other Australia/Japan Living Marine Resources Dispute

Inferences on the Merits of the Southern Bluefin Tuna Arbitration in Light of the Whaling Case

*Andrew Serdy**
Faculty of Business, Law and Art
University of Southampton
A.L.Serdy@soton.ac.uk

Abstract

In 2000 an arbitral tribunal formed under Annex VII to the United Nations Convention on the Law of the Sea denied its own jurisdiction to hear the case brought against Japan by Australia and New Zealand over Japan's unilateral experimental fishing programme for southern bluefin tuna. Despite the criticism the tribunal's reasoning attracted, it was widely supposed that the applicants would have failed on the merits because of the reluctance of international courts and tribunals to delve into scientific matters, as would have been necessary with the dispute's underlying cause being the parties' scientific disagreements regarding both the tuna stock itself and the nature and risks of the programme. In 2014, however, the International Court of Justice showed no such reticence in deciding in Australia's favour the case against Japan's scientific whaling, based partly on flaws it identified in the design of that experiment. Reviewing the evolution in the tuna experiment's design, the propositions it was designed to (dis)prove and the use to which Japan proposed to put that proof, this paper suggests that similar factors were at play in both disputes and that a similar outcome of the tuna case, though not inevitable, would have been amply justified.

Keywords

southern bluefin tuna – dispute – scientific argument – UNCLOS – cooperation

* Certain parts of this contribution have passages in common with the author's recent monograph *The New Entrants Problem in International Fisheries Law* cited in n. 288, for the reason given in the Acknowledgements at page vii of the latter.

 | DOI 10.1163/9789004339453_002

1 Introduction

Readers of the judgment of the International Court of Justice (ICJ) in the *Whaling* case[1] who are familiar with the dispute relating to southern bluefin tuna (SBT) between Australia and New Zealand on one hand and Japan on the other a decade earlier will have experienced a strong sense of *déjà vu*. Although the factual backgrounds of the disputes are only loose parallels of each other, there are numerous and striking similarities in the way Japan conducted its scientific whaling and its experimental fishing for SBT, and a notable continuity of attitudes towards balancing the demands of scientific rigour with other factors militating against this.

The *Southern Bluefin Tuna* arbitration before a tribunal constituted under Annex VII to the United Nations Convention on the Law of the Sea[2] (UNCLOS) and its outcome generated a considerable literature in the following years.[3]

1 *Whaling in the Antarctic (Australia v. Japan: New Zealand intervening)* (Judgment) [2014] ICJ Rep 226.

2 United Nations Convention on the Law of the Sea (opened for signature 10 December 1982, entered into force 16 November 1994) 1833 UNTS 3 (UNCLOS).

3 An incomplete list consists of: Alan Boyle, 'The Southern Bluefin Tuna Arbitration' (2001) 50 *International and Comparative Law Quarterly* 447; David A. Colson and Peggy Hoyle, 'Satisfying the Procedural Prerequisites to the Compulsory Dispute Settlement Mechanisms of the 1982 Law of the Sea Convention: Did the *Southern Bluefin Tuna* Tribunal Get It Right?' (2003) 34 *Ocean Development and International Law* 59; Caroline E. Foster, 'The "Real Dispute" in the Southern Bluefin Tuna Case: a Scientific Dispute?' (2001) 16 *International Journal of Marine and Coastal Law* 571; Mariko Kawano, 'L'affaire du Thon à nageoire bleue et les chevauchements de juridictions internationales' (2003) XLIX *Annuaire Français de Droit International* 516; Barbara Kwiatkowska, 'The Australia and New Zealand v Japan Southern Bluefin Tuna (Jurisdiction and Admissibility) Award of the First Law of the Sea Convention Annex VII Arbitral Tribunal' (2001) 16 *International Journal of Marine and Coastal Law* 239; Barbara Kwiatkowska, 'The *Southern Bluefin Tuna* Arbitral Tribunal Did Get It Right: A Commentary and Reply to the Article by David A. Colson and Dr. Peggy Hoyle' (2003) 34 *Ocean Development and International Law* 369; J.-J. Maguire, 'Southern Bluefin Tuna Dispute' in Myron H. Nordquist and John Norton Moore (eds), *Current Fisheries Issues and the Food and Agriculture Organization of the United Nations* (Martinus Nijhoff Publishers 2000) 201; Bill Mansfield, 'The Southern Bluefin Tuna Arbitration: Comments on Professor Barbara Kwiatkowska's Article' (2001) 16 *International Journal of Marine and Coastal Law* 361; Bill Mansfield, 'Compulsory Dispute Settlement after the Southern Bluefin Tuna Award' in Alex G. Oude Elferink and Donald R. Rothwell (eds), *Oceans Management in the 21st Century: Institutional Frameworks and Responses* (Martinus Nijhoff Publishers 2004) 255; Donald L. Morgan, 'A Practitioner's Critique of the Order Granting Provisional Measures in the Southern Bluefin Tuna Cases' in Myron H. Nordquist and John Norton Moore (eds), *Current Marine Environmental Issues and the International Tribunal for the Law of the Sea* (Martinus Nijhoff Publishers 2001)

It came about because Japan had embarked unilaterally on a programme of experimental fishing in excess of its last agreed national allocation after failing to persuade the applicants, Australia and New Zealand, of the utility of a joint programme of such fishing, which they considered did not meet the standards agreed earlier in the consideration of the concept by the Commission for the Conservation of Southern Bluefin Tuna (CCSBT).[4] The applicants' claim was that this was in breach of Japan's obligations of cooperation with them to take measures required for the conservation and management of the living resources of the high seas, specifically SBT, under Articles 64 and 116–119 of UNCLOS. As subsidiary obligations going to the manner of applying those articles they also invoked Article 300 (the duty to act in good faith and avoid abuse of rights) and the precautionary principle.[5]

173; Donald L. Morgan, 'Implications of the Proliferation of International Legal Fora: The Example of the Southern Bluefin Tuna Cases' (2002) 43 *Harvard International Law Journal* 541; Francisco Orrego Vicuña, 'From the 1893 *Bering Sea Fur Seals Case* to the 1999 *Southern Bluefin Tuna Cases*: A Century of Efforts at Conservation of the Living Resources of the High Seas' (1999) 10 *Yearbook of International Environmental Law* 40; Yoshio Otani 'Quelques réflexions sur la juridiction et la recevabilité vis-à-vis de l'*Affaire du thon à nageoire bleue*' in Nisuke Ando, Edward McWhinney and Rűdiger Wolfrum (eds), *Liber Amicorum Judge Shigeru Oda* (Kluwer Law International 2002) 731; Bernard H. Oxman, 'Complementary Agreements and Compulsory Jurisdiction' (2001) 95 *American Journal of International Law* 277; Jacqueline Peel, 'A Paper Umbrella Which Dissolves in the Rain? The Future for Resolving Fisheries Disputes under *UNCLOS* in the Aftermath of the Southern Bluefin Tuna Arbitration' (2002) 3 *Melbourne Journal of International Law* 53; Tom Polacheck, 'Experimental catches and the precautionary approach: the Southern Bluefin Tuna dispute' (2002) 26 *Marine Policy* 283; Volker Röben, 'The Southern Bluefin Tuna Cases: Re-Regionalization of the Settlement of Law of the Sea Disputes?' (2002) 62 *Zeitschrift für ausländisches öffentliches Recht und Völkerrecht* 61; Cesare Romano, 'The *Southern Bluefin Tuna* Dispute: Hints of a World to Come... Like It or Not' (2001) 32 *Ocean Development and International Law* 313; Tim Stephens, 'The Limits of International Adjudication in International Environmental Law: Another Perspective on the Southern Bluefin Tuna Case' (2004) 19 *International Journal of Marine and Coastal Law* 177; Norio Tanaka, 'Some Observations on the Southern Bluefin Tuna Arbitration Award' (2001) 44 *Japanese Annual of International Law* 9.

4 The CCSBT was established by the Convention for the Conservation of Southern Bluefin Tuna (signed 10 May 1993; entered into force 20 May 1994) 1819 UNTS 359. In 1996, it decided on objectives and principles for the design and implementation of an experimental fishing programme, which required that any programme should be the product of collaboration between the parties, not jeopardise the potential recovery of the parental stock, and be designed to deliver scientifically valid and meaningful results: see infra, (nn. 139–142) and accompanying text.

5 Australia and New Zealand sought by way of relief from the Annex VII Tribunal a declaration that Japan had breached its obligations under arts 64 and 116 to 119 of UNCLOS by "failing

The experimental fishing conducted by Japan in 1998 and 1999 was aimed at reducing one aspect of the uncertainty in the scientific interpretation of catch data on which the parties had been unable to develop an agreed approach, namely the fact that the locations in which fishing took place were not the same from year to year.

Until the *Whaling* case, the fact that the Annex VII Tribunal found that it lacked jurisdiction to determine the claims meant that any systematic analysis of the merits of the *Southern Bluefin Tuna* dispute would have invited the criticism that it was too speculative a basis on which to attempt to draw any firm conclusions. Putting paid to the formerly widespread view that cases in which scientific issues or evidence play a prominent part stand no chance of succeeding because international courts and tribunals cannot cope with scientific arguments, the *Whaling* case and others in the intervening years such as *Pulp Mills*[6] show that this is no longer so. While the International Tribunal for the Law of the Sea (ITLOS) may be proffered as a counterexample, having attracted criticism for disregarding the geological arguments put to it in the *Bay of Bengal* case,[7] that can be readily defended on the footing that the continental shelf provisions of UNCLOS give primacy to geomorphological factors over purely geological ones. This paper makes an attempt at identifying the arguments that Australia and New Zealand might have made had the case reached the merits, concentrating on the Article 119 claim that the experimental fishing was not a measure based on the best scientific evidence available; it does not aim to prove that the claim would have been upheld, only that it was not a foregone conclusion that it would fail.[8] While Japan might have tried

in good faith to co-operate with" them: see Statements of Claim under art. 1 of Annex VII to UNCLOS by which Australia and New Zealand commenced their litigation against Japan, [69 (1)(d)] <https://icsid.worldbank.org/apps/ICSIDWEB/Documents/Statement%20of%20Claim%20of%20Australia%20and%20New%20Zealand.pdf> (accessed 6 August 2015; hereinafter Statement of Claim). Counsel for New Zealand, Ms Geddis, clarified before the tribunal, *Southern Bluefin Tuna Case* (*Australia and New Zealand v. Japan*) (First Round Presentation of Australia and New Zealand, May 8, 2000 (hereinafter Transcript for 8 May)) 168 <https://icsid.worldbank.org/apps/ICSIDWEB/Documents/First%20Round%20Presentation%20of%20Australia%20and%20New%20Zealand_May%208_2000.pdf> accessed 6 August 2015, that this was not an independent allegation of bad faith, a factor that may have been decisive for the outcome: see infra, text at nn. 203–204.

6 *Pulp Mills on the River Uruguay* (*Argentina v. Uruguay*) *Judgment* [2010] ICJ Rep 14.

7 Delimitation of the Maritime Boundary in the Bay of Bengal (Bangladesh/Myanmar) (Judgment of 14 March 2012) [2012] ITLOS Rep 4.

8 As a member of the Australian legal team throughout the *Southern Bluefin Tuna* dispute, the author's recollection is that, beyond some general misgivings expressed on this score, no

to make something by way of counterclaims of Australia's possible contribution to the depletion of the SBT parental biomass before catch restrictions were introduced in 1984, or of potential Australian overcatches in later years through the way in which it accounted for bycatch of SBT by Australian vessels targeting other species, farmed fish and recreational catch,[9] those issues are beyond the paper's scope. In any event, New Zealand's SBT fishery was too small to make New Zealand vulnerable to a similar counterclaim.

2 The *Whaling* Case Judgment

In the *Whaling* case, Australia persuaded the ICJ that Japan's continued pursuit of a large-scale programme of whaling under the Second Phase of its Japanese Whale Research Program under Special Permit in the Antarctic (JARPA II) contravened its obligations under the International Convention for the Regulation of Whaling (ICRW),[10] in particular Article VIII, paragraph 1, whose first sentence reads as follows:

> Notwithstanding anything contained in this Convention any Contracting Government may grant to any of its nationals a special permit authorizing that national to kill, take and treat whales for purposes of scientific research subject to such restrictions as to number and subject to such other conditions as the Contracting Government thinks fit, and the killing, taking, and treating of whales in accordance with the provisions of this Article shall be exempt from the operation of this Convention.

The substantive provisions regulating the taking of whales are found in the Schedule to the ICRW, which by Article V, paragraph 1 may be amended by the International Whaling Commission (IWC) created by Article III, paragraph 1 of the ICRW. This requires a three-quarters majority of votes cast.[11]

---

concerted thought was given to any aspect of the merits in advance of the hearing on Japan's preliminary objections to the tribunal's jurisdiction, other than deciding what relief should be sought in the application instituting proceedings (see Statement of Claim, n. 5 [69]) a necessary step for securing provisional measures from ITLOS, but one which, once taken, immediately yielded priority to that aim.

9 See Andrew Serdy, 'Accounting for Catch in Internationally Managed Fisheries: What Role for State Responsibility?' (2010) 15 *Ocean and Coastal Law Journal* 23, 45–46, 58–65.

10 International Convention for the Regulation of Whaling (signed 2 December 1946; entered into force 10 November 1948) 161 UNTS 72 (ICRW).

11 Ibid. art. III(2).

An amendment becomes binding on a State party unless it presents an objection (which avoids that consequence until the objection is withdrawn).[12]

Specifically, Australia claimed that JARPA II was not a programme for purposes of scientific research within the meaning of Article VIII of ICRW and that Japan had therefore breached three of its obligations under the Schedule: the moratorium setting zero catch limits for the killing of whales from all stocks for commercial purposes (paragraph 10(e)), the prohibition of commercial taking of fin whales in the Southern Ocean Sanctuary (paragraph 7(b)) and the moratorium on the taking, killing or treating of whales, except minke whales, by factory ships or vessels working with such ships (paragraph 10(d)). Japan argued in its defence that none of the provisions invoked by Australia applied to JARPA II, which had been undertaken for purposes of scientific research and was therefore exempted by Article VIII, paragraph 1 from all other provisions of the ICRW.

The case thus turned on whether Japan's whaling conducted under a special permit met the conditions of Article VIII so as not to be subject to the relevant paragraphs of the Schedule. The Court's analysis of this question proceeded in two stages: it did not merely assess whether JARPA II constituted scientific research, but went on to test whether its design and implementation were reasonable in relation to achieving its stated objectives.[13] It contrasted this with resolving matters of scientific or whaling policy, which were beyond its mandate.[14]

With rather odd results, the Court separated into two elements the crucial phrase "for purposes of scientific research",[15] such that, while it found that JARPA II did indeed involve scientific research,[16] the killing, taking and treating of whales pursuant to that programme fell outside Article VIII because these acts were not "for purposes of" scientific research.[17] Having at first resisted it, towards the end of the oral proceedings Japan adopted the view of Australia and New Zealand that the test for this was whether they were "objectively reasonable" or "objectively justifiable" in the sense that they were "supported by coherent reasoning and respectable scientific evidence".[18]

---

12 Ibid. art. V(3).
13 *Whaling in the Antarctic* (n. 1) 254 [67].
14 Ibid. [69].
15 Ibid. 255 [71].
16 Ibid. 267 [127].
17 Ibid. 293 [227].
18 Ibid. 254 [66].

In so doing the Court rejected Australia's definition of the term "scientific research" (on which the ICRW is silent) and considered it unnecessary to put forward an alternative definition of its own.[19] It also expressly avoided passing judgment on the scientific merit or importance of the stated scientific objectives of JARPA II, preferring to assess the purpose of the killing of whales it entailed, and determine whether the design and implementation of JARPA II were the best possible means of achieving those objectives. To this end it examined several elements of the design and implementation of JARPA II, among which were the scale of its use of lethal sampling; the methodology used to select sample sizes; a comparison of the target sample sizes and the actual take; and its timeframe.[20]

JARPA II, in operation since 2005, had four research objectives: monitoring of the Antarctic ecosystem, modelling competition among whale species and future management objectives, elucidation of temporal and spatial changes in stock structure, and improving the management procedure for Antarctic minke whale stocks.[21] It was a long-term research programme contemplated as running indefinitely, but in six-year phases, in an area located within the Southern Ocean Sanctuary established in paragraph 7 (b) of the Schedule.[22] It would consist of a mixture of lethal sampling of 850 Antarctic minke whales, 50 fin whales and 50 humpback whales, as well as non-lethal methods, namely biopsy sampling, satellite tagging and whale sighting surveys.[23] Based on current abundance estimates, Japan expected the planned take of each species to be too small to have any negative effect on the stocks of these species.[24]

The Court examined at some length whether the design and implementation of JARPA II were reasonable in relation to achieving its stated objectives. Although it found no basis to conclude that the use of lethal methods was unreasonable in itself,[25] it was more troubled by Japan's disregard of the IWC resolutions and Guidelines of its Scientific Committee calling upon members to take into account whether research objectives could be achieved using non-lethal methods, noting the expert evidence led by Australia on significant advances in a wide range of non-lethal research techniques over the past

19 Ibid. 258 [86].
20 Ibid. [88].
21 Ibid. 263–264 [109], [113].
22 Ibid. 266 [119]–[120].
23 Ibid. 266–267 [121], [123], [124].
24 Ibid. 267 [126].
25 Ibid. 269 [135].

20 years with potential application to JARPA II's stated objectives.[26] In particular, Japan had not considered the feasibility or practicability of non-lethal methods either in initially setting the JARPA II sample sizes or in their maintenance in later years, or of combining a smaller lethal take with an increase in non-lethal sampling as a means to achieve those objectives.[27]

Turning to the scale of the use of lethal methods in JARPA II, the Court noted that, despite the considerable overlap between the subjects, objectives, and methods of JARPA II and the programme it succeeded (JARPA), the sample size for minke whales (850 ± 10 per cent) was nearly double the minke whale sample size for the last years of the predecessor programme, which moreover did not include any lethal sampling of fin and humpback whales.[28] These similarities cast doubt on the objectives relating to ecosystem monitoring and multi-species competition as justification for this significant quantitative and qualitative expansion in lethal sampling.[29] On the basis of the stress laid by Japan on the need for continuity between the two programmes, the Court also questioned whether the sample sizes and launch date for JARPA II were motivated by purely scientific considerations.[30]

The Court paid close attention over several dozen paragraphs[31] to how the species-specific sample sizes were determined, taking care to indicate that it was not in a position to conclude whether one particular value for a given variable had scientific advantages over another; rather, its focus was solely on whether the evidence supported the conclusion that the sample sizes were reasonable for achieving the stated objectives of JARPA II.[32] This evidence the Court found largely unconvincing. Notably, it found that the initial sample sizes for humpback and fin whales (50 of each) were too small to detect a particular rate of change in one parameter, for which 131 whales of each species would have been needed, but Japan had offered no evidence as to whether its researchers had decided to accept a lower level of accuracy or instead adjusted the rate of change to be detected by targeting fewer whales.[33] Equally significant was the gap between the JARPA II target sample sizes and the actual number of whales killed: only 18 fin whales over the first seven seasons and no

26 Ibid. 269–270 [137].
27 Ibid. 271 [141], [144].
28 Ibid. 272 [148].
29 Ibid. 274 [153].
30 Ibid. [154]–[156].
31 Ibid. 275–286 [157]–[198].
32 Ibid. 278 [172].
33 Ibid. 280 [179].

humpbacks at all, together with a drop in the take of minke whales from 853 during the 2005–06 season, approximately 450 in the next few seasons, 170 in 2010–11 and 103 in 2012–13.[34]

Further, despite this persistent difference between the implementation of JARPA II and its design, the fact that Japan had made no changes to the objectives and target sample sizes, coupled with its contention that JARPA II could yield meaningful scientific results from the far smaller actual take, further undermined the proposition that it was a programme for purposes of scientific research. To the contrary, for the Court this was evidence that the target sample sizes were larger than was reasonable for achieving JARPA II's stated objectives.[35] Japan's admission that the actual take of fin and humpback whales was mostly, if not wholly, a function of political and logistical considerations served also to weaken the relationship between JARPA II's research objectives and the decision to engage in the lethal sampling of minke whales on a relatively large scale.[36]

Other factors supporting this conclusion were that a "time frame with intermediary targets" would have been more appropriate,[37] as the IWC Scientific Committee's Guidelines indicated,[38] and the fact that, even though the first six-year phase of JARPA II (2005–06 to 2010–11) had already been completed, it had so far resulted in only two peer-reviewed papers.[39] Moreover, there was little evidence of the co-operation between JARPA II and other domestic and international research institutions that could have been expected given its focus on the Antarctic ecosystem and environmental changes in the region.[40]

---

34 Ibid. 286 [199], [201]–[202].

35 Ibid. 289–290 [209]–[212].

36 Ibid. 286 [203].

37 Ibid. 290 [216].

38 In 1950, the IWC created a Scientific Committee whose tasks under paragraph 30 of the Schedule include reviewing and commenting on special permits before they are issued by States parties to their nationals for purposes of scientific research under art. VIII(1) of the ICRW: ibid. 248 [47]. Since the mid-1980s, the Scientific Committee has conducted its review of special permits on the basis of guidelines issued or endorsed by the IWC. When JARPA II was proposed in 2005, the applicable guidelines had been collected in a document titled 'Annex Y: Guidelines for the Review of Scientific Permit Proposals'. The current guidelines are found in a newer document, 'Annex P: Process for the Review of Special Permit Proposals and Research Results from Existing and Completed Permits'. See ibid. 252 [58] and 257 [84].

39 Ibid. 291 [219].

40 Ibid. 292 [222].

Lastly, Japan had not offered any explanation for why the research periods differed as between species.[41]

All of this led the Court to conclude that, while JARPA II involved activities that could broadly be characterised as scientific research, the evidence did not establish that its design and implementation were reasonable in relation to achieving its stated objectives. Accordingly, the special permits granted by Japan for the killing, taking and treating of whales in connection with JARPA II were not "for purposes of scientific research" pursuant to Article VIII, paragraph 1 of the ICRW.[42]

It followed that Japan by its issuance of JARPA II permits for 850 minke whales, 50 fin whales and 50 humpback whales had set catch limits above zero for these species, which was not in conformity with its paragraph 10(e) obligation in each of the years in question.[43] The location of some of these catches in the Southern Ocean Sanctuary (only in respect of fin whales, since Japan had objected to its application to minke whales, and no humpback whales were taken), and the use of the factory ship *Nisshin Maru* to this end brought about violations of paragraphs 7(b) and 10(d) of the Schedule respectively.[44] As JARPA II would otherwise have continued indefinitely, the Court ordered Japan to revoke any extant authorisation, permit or licence to kill, take or treat whales under that programme and to refrain from granting any further permits under Article VIII, paragraph 1 of the ICRW in pursuance of it.[45]

All of the foregoing suggests that the Court had no trouble coping with the scientific arguments put to it. With this in mind, the remainder of this paper canvasses what an international court or tribunal might have made of the merits of the *Southern Bluefin Tuna* dispute if it had arisen closer to the present day.

3 Southern Bluefin Tuna: The Fish

SBT is one of the seven of the tuna species considered by the FAO as principal market tuna species because of their global economic importance and their international trade for canning and sashimi, raw fish being regarded as a deli-

41 Ibid. 293 [226].

42 Ibid. [227], 299 [247(2)] (Judges Owada, Abraham, Bennouna and Yusuf dissenting).

43 Ibid. 295 [231], 299 [247(3)] (Judges Owada, Abraham, Bennouna and Yusuf dissenting).

44 Ibid. 295–296 [232]–[233], 299 [247(4) and (5)] (Judges Owada, Abraham, Bennouna and Yusuf dissenting).

45 Ibid. 298 [245], 300 [247(7)] (Judges Owada, Abraham, Bennouna and Yusuf dissenting).

cacy in Japan and more recently also in other countries.[46] SBT, though of only minor significance in terms of volume,[47] realises high prices in the sashimi trade. Although prices for bluefin species have fallen, some years ago being generally in the range US$30–100, the highest quality individual fish can sometimes fetch prices above US$500 per kilogram if they are properly handled after capture.[48]

Along with all but three of the 15 tuna species thought to exist, SBT is listed in Annex I to UNCLOS as a highly migratory species to which Article 64 applies.[49] The intention appears to have been to include all known tunas in Annex I,[50] but was imperfectly executed. It ranges widely across the high seas regions of the southern temperate oceans, particularly between 30° and 45° South, but also traversing the exclusive economic zone (EEZ) and territorial sea of several States including Australia, New Zealand, Indonesia and South Africa. This makes it the most widely dispersed stock of all the tunas.[51]

---

46 Jacek Majkowski and others, 'Tuna and Tuna-like Species' in FAO, *Review of the state of world marine fishery resources* (FAO Fisheries and Aquaculture Technical Paper 569, FAO 2011) 227, 227 <http://www.fao.org/docrep/015/i2389e/i2389e.pdf>. Six of the seven principal market tunas are of the genus *Thunnus*: albacore (*T. alalunga*), bigeye tuna (*T. obesus*), Atlantic bluefin tuna (*T. thynnus*), Pacific bluefin tuna (*T. orientalis*), Southern bluefin tuna (*T. maccoyii*) and yellowfin tuna (*T. albacares*); the seventh is skipjack tuna (*Katsuwonus pelamis*).

47 Ibid. 234.

48 Ibid. 227.

49 Annex I lists only 11 species, but was drafted before the distinction between Atlantic and Pacific bluefin tuna was recognised, thus treating the two species as one, bluefin tuna *simpliciter* (*T. thynnus*): see Andrew Serdy, 'One fin, two fins, red fins, bluefins: some problems of nomenclature and taxonomy affecting legal instruments governing tuna and other highly migratory species' (2004) 28 *Marine Policy* 235, 235–236.

50 A monograph by the deputy head of the Spanish delegation to the Conference states that the highly migratory species include "all the varieties of tuna": José Antonio de Yturriaga, *The International Regime of Fisheries: From UNCLOS 1982 to the Presential Sea* (Martinus Nijhoff Publishers 1997) 128.

51 Tom Polacheck, 'An overview of interaction issues among the fisheries for Southern Bluefin Tuna' in Richard S. Shomura, Jacek Majkowski and Sarah Langi (eds), *Interactions of Pacific Tuna Fisheries: Proceedings of the First FAO Expert Consultation on Interactions of Pacific Tuna Fisheries* (3–11 December 1991) vol. 1 (FAO Technical Paper 336/1, FAO 1994) 264, 264; *Report on Biology, Stock Status and Management of Southern Bluefin Tuna* (Attachment 6 to CCSBT, *Report of the Extended Scientific Committee for the Ninth Meeting of the Scientific Committee* (13–16 September 2004) (hereinafter CCSBT-ESC Report 2004; Appendix 2 to CCSBT, *Report of the Ninth Meeting of the Scientific Committee* (13–16 September 2004) 1 <http://www.ccsbt.org/userfiles/file/docs_english/meetings/meeting_reports/ccsbt_11/report_of_sc9.pdf> accessed 6 August 2015)).

It is generally accepted that the global population of SBT comprises a single stock, with a single spawning ground situated in the tropical waters south of the Indonesian islands from Java to beyond Sumba and extending at its eastern end well into the Australian EEZ in the vicinity of the Western Australian port of Broome.[52] By maturity, most fish have dispersed into the deeper waters of the Indian, South Atlantic and south-west Pacific Oceans. SBT tend to school in order to feed, making them susceptible to capture by longlining.[53]

There is some uncertainty as to when it reaches maturity and can begin spawning but the mean age for maturity is thought likely to be 11 or 12 years.[54] Precisely because of its longevity coupled with life-long exposure to fishing pressure, however, SBT has been prone to overexploitation. A consequence of this is that, once its numbers are depleted, they would not be expected to recover until some time after reduction in fishing effort.[55]

---

52 A. Caton, K. McLoughlin and M.J. Williams, *Southern bluefin tuna: scientific background to the debate* (Australian Government Publishing Service 1990) 7 (map), 10.

53 John O.S. Kennedy, Luke Davies and Anthony Cox, *Joint Rent Maximisation and Open Access Competition in the Southern Bluefin Tuna Fishery* (Australian Bureau of Agricultural and Resource Economics Conference Paper 99.1) 3 <http://data.daff.gov.au/brs/data/warehouse/pe_abarebrs99000382/PR11266.pdf> accessed 6 August 2015.

54 This was an element in the scientific and subsequently legal controversy of the 1990s: see infra (n. 283) and accompanying text and Appendix, text between (nn. 320 and 321). The report of the 1994 scientific meeting defined the parental stock as fish of age 8 and older, on the basis that although some 7-year-old fish were mature while not all 9-year-old fish were mature, 50% of SBT were thought to be mature at age 8: *Report of the thirteenth meeting of Australian, Japanese and New Zealand Scientists on southern bluefin tuna: Report to Management* (20–29 April 1994) (hereinafter 13th Trilateral Scientific Meeting Report; unpublished, copy held by author extracted from files of the former Australian Government Department of Primary Industries and Energy (hereinafter DPIE)), 8 (Appendix 1, 'Status of the stock and fishery indicators', [4]). In 2001 an independent advisory panel found that the formerly agreed estimated age of 8, on which Japan had continued to insist in the face of newer evidence to the contrary, was "unlikely" and recommended the use of alternative hypotheses, ages 10 and 12: CCSBT, *Report of the Second Meeting of the Stock Assessment Group* (19–28 August 2001) (hereinafter CCSBT-SAG2 Report) 4 [14] <http://www.ccsbt.org/userfiles/file/docs_english/meetings/meeting_reports/ccsbt_08/report_of_sag2.pdf> accessed 6 August 2015; see also Attachment 4, 'Working paper for maturity age group'. By implication the scientists settled on age 10 the following year, defining "adult mortality" by reference to that age in CCSBT, *Report of the Third Meeting of the Stock Assessment Group* (3–7 September 2002) (hereinafter CCSBT-SAG3 Report) 5 [22] <http://www.ccsbt.org/userfiles/file/docs_english/meetings/meeting_reports/ccsbt_09/report_of_sag3_short.pdf>.

55 Majkowski and others (n. 46) 235; see also infra Appendix. Stocks of shorter-lived species, including most other species of tuna, recover much faster from depletion. For example,

4 Historical Overview of the Japanese SBT Fishery

Significant commercial harvest of SBT began in the early 1950s, the Australian and Japanese fisheries developing roughly at the same time but independently of each other. The highest catch of over 81,000 tonnes occurred in 1961. In 2004 it was estimated that 73% of the catch had been taken in the Indian Ocean, 21% in the Pacific and 6% in the Atlantic, mostly off the southern tip of Africa.[56]

Offshore fishing operations began to receive official encouragement from the Japanese authorities in the 1880s, so that there would be adequate food supplies in years when the harvest of rice and other crops failed.[57] Longline fishing developed in the late nineteenth century,[58] but this was initially of little significance for tunas, for which demand was low, the fishing for it being conducted close to Japan's coast.[59] Severe food shortages recurred in the 1940s and into the 1950s because of the loss of manpower and of 60 per cent of the Japanese tuna fleet in the Second World War. The Government therefore took measures to stimulate food production, including bringing the tuna fisheries under military control in 1942,[60] and after 1945 fisheries production recovered rapidly as the Government promoted expansion of fishing grounds and subsidised vessel construction.[61]

After Japan's surrender the occupation authorities started by completely prohibiting the movement of Japanese fishing vessels, but later progressively expanded the high seas areas in which they were permitted to fish.[62] By 1950

---

yellowfin and skipjack tunas, which are caught primarily for the canned tuna market, reach maturity at a young age (less than 2 years) and are highly productive: Caton, McLoughlin and Williams (n. 52) 10.

56 Attachment 6 to CCSBT-ESC Report 2004 (n. 51) 205.

57 Norio Fujinami, 'Development of Japan's Tuna Fisheries' in David J. Doulman (ed), *Tuna Issues and Perspectives in the Pacific Islands Region* (Pacific Islands Development Program 1987) 57.

58 Ibid. 58.

59 According to a chronicler of life in old Edo, tuna was formerly held in such low esteem that not even the poor would eat it, possibly because its warm flesh spoils quickly: Steve Williams, 'Understanding Japanese seafood markets' (1992) 51(2) *Australian Fisheries* 32, 35n.

60 Yoshiaki Matsuda, 'Postwar Development and Expansion of Japan's Tuna Fishery' in Doulman (ed) (n. 57) 71.

61 Fujinami (n. 57) 58.

62 Matsuda (n. 60) 72. Harry N. Scheiber, 'Origins of the Abstention Doctrine in Ocean Law: Japanese-U.S. Relations and the Pacific Fisheries, 1937–1958' (1989) 16 *Ecology Law Quarterly* 23, 40ff describes how, as part of the general programme of economic revival, the resumption of fisheries was actively encouraged by occupation authorities, not only

the fish resources of the area within the outermost of these "MacArthur lines" were fully exploited. Once the last spatial restriction was removed in 1952 when the peace treaty entered into force,[63] Japanese policy moved to encourage the development of distant-water fishing, including transfer of licences from the fully exploited offshore fisheries and encouraging construction of larger vessels. Catches of tuna rose quickly. By 1964 Japan's distant-water tuna fleet was operating throughout the world, licensed vessel numbers reaching 3,000 in 1965, of which about 1,300 pole-and-line and longline vessels had no restrictions on their area of operation.[64] Thanks to progress in freezing technology, in the late 1960s it became possible for the fleet to shift from supplying canneries to producing sashimi-grade tuna.[65] Soon afterwards, however, the longline fleet became uneconomic at its then size owing to a number of factors: rising fuel and (because of Japan's rapid post-war economic growth) crewing costs, declining catch rates and changing food consumption patterns. A specifically legal factor exacerbating this trend was the extension of coastal State jurisdiction under the new concept of the EEZ,[66] resulting in closure of many areas to fishing and payments for access to those areas that remained open.

The first Japanese vessels to fish for SBT began doing so in 1952, south of Java on the SBT spawning grounds, as soon as the last MacArthur line was lifted.[67] Eventually the Japanese longline fishery expanded into most of the southern Indian, Pacific and Atlantic Oceans, extending at its peak from 10°E to 170°W with concentrations off Western Australia, South Africa, Tasmania and New

---

to promote self-sufficiency in food, but later also, with the onset of the Cold War, for geopolitical reasons.

63 Treaty of Peace with Japan (signed 8 September 1951, entered into force 28 April 1952) 136 UNTS 45.

64 Fujinami (n. 57) 58–59. Licensing requirements applied to longliners and pole-and-line vessels of more than 20 gross registered tons (GRT) and purse seiners above 40 GRT. Other than longliners and pole-and-line vessels of more than 120 GRT, the Government restricted tuna vessels' area of operation in order to ensure their safety, avoid concentration of fishing effort and assist orderly marketing of the catch: ibid. 65.

65 Ibid. 60–61. This change occurred in 1972 for SBT, which until then had been used for teriyaki and fish sausages: Industries Assistance Commission, *Report on Southern Bluefin Tuna* (Australian Government Publishing Service 1984) 87 (evidence of the Executive Director of the Federation of Japan Tuna Fisheries Co-operative Associations).

66 According to Matsuda (n. 60) 86, in 1977 48% of Japan's catch of tunas other than skipjack was taken within 200 nautical miles of other States' coasts.

67 Matsuda (n. 60) 84.

Zealand.[68] The main longline fishing grounds shifted seasonally in line with changes in ocean conditions. With the introduction of monthly sea surface temperature charts as an aid in locating fish, the efficiency of fishing operations rose.[69] In both 1960 and 1961 catches in excess of 75,000 tonnes were recorded, after which the catch fluctuated for the next decade around 40,000 tonnes with occasional spikes to 50,000 or 60,000 tonnes.[70] By 1970 the fleet was 1,200 vessels strong, including over 300 specialist SBT boats that shifted between the Indian and South Atlantic Oceans depending on where fishing conditions were better,[71] but the decade that followed saw further decline in annual catch to around 30,000 tonnes and falling economic returns, together leading to a reduction in vessel numbers by 20 per cent.

The decline in the catch rate of SBT (catch per unit of effort, hereinafter CPUE) was even greater. From 1961 to 1987 Japanese effort rose from 30 million to 110 million hooks, meaning that the catch rate declined by around 95% over that period.[72] The decline in abundance may have been greater still once the

68 Ibid. 78; Caton, McLoughlin and Williams (n. 52) 13; B.B. Collette and C.E. Nauen, *FAO Species Catalogue Vol. 2: Scombrids of the World: An Annotated and Illustrated Catalogue of Tunas, Mackerels, Bonitos and Related Species Known to Date* (FAO Fisheries Synopsis No 125 vol. 2; FAO 1983) 88.

69 Ibid.

70 'Global Reported Catch by Flag' (Attachment 4 to CCSBT, *Report of The Extended Scientific Committee for the Twentieth Meeting of the Scientific Committee* (1–5 September 2015) (hereinafter CCSBT-ESC Report 2015; Appendix 2 to CCSBT, *Report of the Twentieth Meeting of the Scientific Committee* (5 September 2015) <https://www.ccsbt.org/sites/ccsbt.org/files/userfiles/file/docs_english/meetings/meeting_reports/ccsbt_22/report_of_SC20.pdf> accessed 31 March 2016).

71 Caton, McLoughlin and Williams (n. 52) 13; *Report of the Standing Committee on Research and Statistics (SCRS)* (5–10 November 1981) (Annex 8 to ICCAT, *Proceedings of the Seventh Regular Meeting of the Commission*), in International Commission for the Conservation of Atlantic Tunas (hereinafter ICCAT), *Report for biennial period* 1980–81 Part II (1981) 108, 138. A significant part of Japan's catch continues to be taken in the south-east Atlantic in some years, e.g. 1,205 tonnes in 2000: 'National Report of Japan' in ICCAT, *Report for biennial period* 2002–03 Part II (2003) vol. 3, 53, 61 (Table 3). The table shows that 2,506 tonnes of SBT were caught in the Atlantic in 1981, and over 1,000 tonnes in ten of the next 20 years, the highest figure being 1,688 tonnes in 1993, and the lowest 301 tonnes in 1997.

72 According to Caton, McLoughlin and Williams (n. 52) 17, CPUE is frequently used as a surrogate measure of fish abundance, although it is not necessarily directly proportional to it. The Japanese longline fishery expresses its CPUE as the number of SBT caught per 1000 hooks set.

increasing efficiency of the fleet is taken into account.[73] In 1971 the Japanese fleet voluntarily adopted a seasonal closure south of the spawning area in order to protect migrating spawning fish, reacting to the findings of Japanese scientists that the mean density of spawning adults had been decreasing since the early 1960s and on the spawning ground itself had by 1967–68 fallen by nine tenths.[74] Similar closures followed in areas off southern New South Wales, South Australia and South Africa where immature fish were predominant, in line with the scientists' views that the exploitation of younger fish was too high.[75] Of greater concern to Japanese scientists, on the feeding grounds effort continued to grow during the 1970s, though there too catch rates declined.[76] The closures were only partly effective: recruitment to the surface fisheries was maintained, but declined to the longline fishery harvesting fish four years and older.[77]

In response, the Japanese vessels began to concentrate their effort spatially, fishing only the historically more productive locations, an additional factor tending to make raw CPUE figures overestimate actual abundance. The Japanese industry was not able to fill its quota despite the marked downward

---

73 Ibid.; on the increase in efficiency over time see also the oral submissions of Professor Crawford, Counsel for Australia, to ITLOS on 18 August 1999: *Southern Bluefin Tuna Cases* (*New Zealand v. Japan; Australia v. Japan*) (Request for provisional measures) ITLOS doc ITLOS/PV.99/21/Rev.2 (hereinafter Transcript for afternoon of 18 August), 15 <https://www.itlos.org/fileadmin/itlos/documents/cases/case_no_3_4/VRE180899pm.corr-rev2.pdf> accessed 3 August 2015.

74 Caton, McLoughlin and Williams (n. 52) 19, 31. According to ICCAT, *Report of the Standing Committee on Research and Statistics* (SCRS) (9–15 November 1977) (Annex 9 to ICCAT, *Proceedings of the Fifth Regular Meeting of the Commission* (16–22 November 1977) (hereinafter ICCAT5 Report), in ICCAT, *Report for biennial period* 1976–77 Part II (1977) 98, 150, the spawning stock "may have been reduced to 10–20% of the earlier levels." This was repeated in ICCAT, *Report of the Standing Committee on Research and Statistics* (SCRS) (8–14 November 1978) (Annex 6 to ICCAT, *Proceedings of the First Special Meeting of the Commission* (15–21 November 1978), in ICCAT, *Report for biennial period* 1978–79 Part I (1978) 93, 157. Japan explained the aim of the measures to ICCAT as avoiding catching fish younger than 7, though longlines caught fish aged as young as 4: Annex 9 to ICCAT5 Report (this n.) 150.

75 Caton, McLoughlin and Williams (n. 52) 31; Collette and Nauen (n. 68) 88.

76 Caton, McLoughlin and Williams (n. 52) 19, 31.

77 Annex 9 to ICCAT5 Report (n. 74) 150. Recruitment is defined as the quantity of fish added to the fishery each year by becoming vulnerable to the fishing gear through growth or migration into the fishing area: 'Definition of Technical Terms' (Appendix 12 to ICCAT, *Report of the Standing Committee on Research and Statistics* (SCRS) (9–13 October, 1995), in ICCAT, *Report for biennial period* 1994–95 Part II (1995) vol. 2, 158, 162.

trend in the quota allocations to Japan when they commenced in 1985. Data from the Japanese longline fishery within the Australian Fishing Zone (AFZ) also showed hook rates down in the late 1980s to only half their level of a decade earlier. Although a 50 per cent increase was reported in the AFZ in the later years of the decade, which may have been a result of the reduction in fishing mortality from the Australian surface fishery as a consequence of the sharply reduced Australian quotas,[78] the evidence was inconclusive, since no similar improvement in the numbers of small fish occurred in the New Zealand fishery;[79] oceanographic conditions of the time or behavioural changes owing to the reduced abundance cannot be ruled out as causes.[80]

5 First Interactions

In the late 1960s Japanese scientists began to express concern in the Indo-Pacific Fisheries Council[81] (now the Asia-Pacific Fishery Commission) and later the Indian Ocean Fisheries Commission[82] and both bodies' tuna management committees that the Australian fishery developing on juvenile SBT was an obstacle to the aim of increasing the size of fish caught in the Japanese longline fishery.[83] After several years of inaction, a "special southern bluefin tuna working party" of Australian and Japanese scientists was convened at Japan's Far Seas Fisheries Research Laboratory in June 1975,[84] which led the

78 Caton, McLoughlin and Williams (n. 52) 20.

79 Ibid. 22.

80 Ibid.; see also the oral submissions of Professor Crawford, Counsel for Australia, to ITLOS, Transcript for afternoon of 18 August (n. 73) 21.

81 Established by the Agreement for the Establishment of the Indo-Pacific Fisheries Council (signed 26 February 1948, entry into force 9 November 1948) 120 UNTS 59.

82 Established by Resolution 2/48 of the FAO Council at its Forty-eighth Session in September 1967, abolished by Resolution 1/116 of the same body at its Hundred and Sixteenth Session in June 1999, reprinted in FAO doc CL 116/REP, *Report of the Hundred and Sixteenth Session of the Council* (14–19 June 1999) [124] <http://www.fao.org/docrep/X2372e/X2372e02.htm> accessed 22 August 2015. The Commission's mandate covered all living marine resources in the Indian Ocean and adjacent seas excluding the Antarctic area: J.J. Kambona and S.H. Marashi, *Process for the Establishment of the Indian Ocean Tuna Commission* (FAO Fisheries Circular No 913; FAO 1996) 2 [5].

83 Caton, McLoughlin and Williams (n. 52) 18, 31.

84 S. Kume, 'Japanese Fisheries and Research Activities on Tunas and Tuna-Like Fisheries in the Atlantic Ocean, 1973–1975' in ICCAT, *Report for biennial period* 1974–75 Part II (1975) 187, 190.

following year to a cooperative study by Australian and Japanese scientists.[85] A joint Working Group on Stock Assessment of the two commissions concluded that longline fishing intensity was high and that the stock was being heavily exploited.[86]

A major change in SBT management arrangements became necessary in the second half of the 1970s when some of the high seas areas in which the Japanese fishery had been prosecuted came under expanded coastal State jurisdiction during the Third United Nations Conference on the Law of the Sea. Even before this culminated in Part V of UNCLOS on the EEZ, many States around this time either declared full EEZs out to 200 nautical miles from their baselines, or claimed exclusive fisheries jurisdiction to the same distance. The declaration of the New Zealand EEZ in 1977,[87] followed in 1979 by Australia's proclamation of the AFZ,[88] meant that the Japanese SBT fleet could henceforth operate in those zones only by agreement of the coastal States. In 1978 Japan negotiated with New Zealand a treaty providing for access of Japanese longliners to the New Zealand EEZ for four years, subsequently extended by a series of exchanges of notes until 1997.[89] A similar instrument was concluded in 1979 between Australia and Japan, along with the first of a long series of annual Subsidiary Agreements setting out the terms on which Japanese longline fishing vessels could fish for tuna in the AFZ.[90]

85 ICCAT, *Report of the Standing Committee on Research and Statistics* (Annex 8 to ICCAT, *Proceedings of the Fourth Regular Meeting of the Council*), in ICCAT, *Report for biennial period* 1976–77 Part I (1976) 69, 103.

86 Ibid. 102.

87 Territorial Sea and Exclusive Economic Zone Act 1977 (NZ), s 9.

88 Commonwealth of Australia, Gazette S189 (26 September 1979), Schedule, taking effect on 1 November 1979. The legislative authority for this proclamation was created a year earlier by *Fisheries Amendment Act 1978* (Cth), ss 3, 6.

89 Agreement on Fisheries between the Government of New Zealand and the Government of Japan (signed 1 September 1978, entry into force 1 September 1978) 1167 UNTS 441; extended by exchanges of Notes of 26 May 1982 (for two years—1324 UNTS 410), 21 September 1984 (for two years—1676 UNTS 553), 23 September 1986 (for four years—1937 UNTS 403), 26 September 1990 (for two years—1937 UNTS 403), 30 September 1992 (for two years—1937 UNTS 403), 30 September 1994 (for two years—1937 UNTS 403) and 23 September 1996 (for one year—1950 UNTS 402).

90 Agreement on Fisheries between the Government of Australia and the Government of Japan and Subsidiary Agreement between the Government of Australia and the Government of Japan concerning Japanese Tuna Long-Line Fishing (signed 17 October 1979, entry into force 1 November 1979) 1217 UNTS 3 (Head Agreement), 19 (Subsidiary Agreement). The agreements entered into force on 1 November 1979, the day on which the AFZ came into effect (n. 88). A further 17 subsidiary agreements covering individual

6 The State of the Stock

Australian scientists first informed their government in 1979 that the global SBT fishery was fully exploited, in other words any additional effort would yield little if any extra catch, since any further increase in exploitation by Australian fishermen would reduce the Japanese catch, and *vice versa*.[91] In 1981 the Commonwealth Scientific and Industrial Research Organisation (CSIRO) gave further warning of the extent of decline in the SBT parental biomass and the high risk to recruitment this entailed,[92] prompting Australia to approach Japan and New Zealand about establishing a tripartite scientific review of the state of the SBT stock and exploring how the fishery might be managed globally.[93] The three States began annual trilateral scientific and management discussions in 1982.[94]

With papers from Australia and New Zealand both showing that parental biomass would continue to decline unless the level of fishing pressure could be reduced,[95] the scientists agreed that "it would be wise to take steps now to stabilise the stock at approximately present levels and to have in place

fishing seasons were negotiated in most of the succeeding years up to 1997, all entitled Subsidiary Agreement concerning Japanese Tuna Long-Line Fishing: (i) (signed 30 October 1980, entry into force 30 October 1980) 1217 UNTS 40; (ii) (signed 29 October 1981, entry into force 1 November 1981) 1342 UNTS 3; (iii) (signed 28 October 1982, entry into force 1 November 1982) 1342 UNTS 41; (iv) (signed 31 October 1983, entry into force 1 November 1983) 1424 UNTS 85; (v) (signed 30 October 1984, entry into force 1 November 1984) 1426 UNTS 29; (vi) (signed 31 October 1985, entry into force 1 November 1985) 1430 UNTS 9 (title page) and 22 (text); (vii) (signed 30 October 1986, entry into force 1 November 1986) 1459 UNTS 197; (viii) (signed 29 October 1987, entry into force 1 November 1987) 1487 UNTS 115; (ix) (signed 27 October 1988, entry into force 1 November 1988) 1536 UNTS 331; (x) (signed 15 December 1989, entry into force 15 December 1989) 1573 UNTS 3; (xi) (signed 30 November 1990, entry into force 30 November 1990) 1598 UNTS 341; (xii) (signed 10 December 1991, entry into force 10 December 1991) 1680 UNTS 407; (xiii) (signed 21 December 1992, entry into force 21 December 1992) 1736 UNTS 115; (xiv) (signed 24 December 1993, entry into force 24 December 1993) 1770 UNTS 457; (xv) (signed 21 December 1994, entry into force 21 December 1994) 1889 UNTS 191; (xvi) (signed 4 June 1996, entry into force 4 June 1996) 1945 UNTS 275; (xvii) (signed 4 June 1997, entry into force 4 June 1997) 2007 UNTS 43.

91 Caton, McLoughlin and Williams (n. 52) 18; Industries Assistance Commission (n. 65) 10.

92 For the definition of this term see (n. 77).

93 Caton, McLoughlin and Williams (n. 52) 32.

94 Ibid. 26, 31.

95 *Record of Australia/Japan/New Zealand Scientific Discussions on Southern Bluefin Tuna* (13–15 December 1982) in Bureau of Rural Resources, *Reports of the Trilateral Scientific Discussions among Australia, Japan and New Zealand on Southern Bluefin Tuna 1982–1991*

mechanisms to further reduce the catches should recruitment begin to falter."[96] The three States held their first management meeting in Wellington in 1982 and agreed that the eventual management framework should be binding, but did not immediately set catch limits, confining themselves to indicating their concern at the decline of the stock.[97]

Thereafter the pattern of scientific meetings followed by management ones continued on a roughly annual cycle, the three States taking it in turn to host the gatherings.[98] The 1982 meeting confirmed the earlier CSIRO estimates of the decline in the parental biomass and concluded that the stock had been fished down to possibly less than a third of its original level. The 1983 meeting predicted unavoidable further decline in the short term, the participants concluding that urgent steps needed to be taken to avert parental biomass falling significantly below the 1980 level, beyond which they considered there was significant risk that satisfactory numbers of recruits could not be produced.[99] The next year they warned that the longer remedial action was deferred, the more severe would be the eventual catch reductions required to keep the spawning stock at a satisfactory level.[100]

In 1985 the scientists expressed the view that managers should in the short term adopt a conservative approach to the SBT fishery and by 1987 they regarded this as imperative, warning that there was risk associated with maintaining the then catch limits.[101] For the fishery as a whole CPUE fell by half from 1983 to 1986 and further still in 1987.[102] Participants at the 1988 meeting concluded that even by 1979 the parental biomass may have been reduced to too low a level, so it was vital to prevent any further decline. To achieve this and safeguard the long-term viability of the fishery, there would need to be substantial further reductions in catches until the parental biomass and the recruitment from it had demonstrably recovered to much higher levels.

(Working Paper No. WP/10/92; Bureau of Rural Resources, 1992) (hereinafter Trilateral Scientific Meeting Reports Compendium) 1, 3.

96 Ibid. 2.

97 Cable No 4569 of 23 December 1982 from the New Zealand High Commission in Canberra to the Ministry of Foreign Affairs in Wellington, in New Zealand Ministry of Foreign Affairs file 40/12/10 'New Zealand Affairs: Economic Relations—Japan—Fishing', Part 22.

98 Trilateral Scientific Meeting Reports Compendium (n. 95) iii.

99 Caton, McLoughlin and Williams (n. 52) 26.

100 *Report of the Third Tripartite Scientific Meeting on Southern Bluefin Tuna* (28 May 1984) in Trilateral Scientific Meeting Reports Compendium (n. 95) 11, 16.

101 Caton, McLoughlin and Williams (n. 52) 27.

102 Ibid. 19.

The only safe catch was zero, but even with zero catch biological mechanisms preventing a recovery could not be ruled out.[103]

7 The First Quotas and Their Subsequent Reduction

The first total allowable catch (TAC) for SBT came about in 1985, when Australia, Japan and New Zealand entered into a voluntary trilateral arrangement to this end, under which they negotiated quotas each year drawing on the advice from the trilateral scientific meetings. Australia's quota was 14,500 tonnes, that of Japan was 23,150 tonnes, while as a relative newcomer to the fishery New Zealand had a quota of 1,000 tonnes expressed as being for development purposes, representing several multiples of its actual catch. In 1986 and 1987 Australia and Japan left their 1985 quotas nominally unchanged but agreed that their catches would not exceed 11,500 tonnes and 19,500 tonnes respectively.[104]

These cuts did not succeed in stopping the stock's decline. The conclusion from the 1988 scientific meeting was that the only safe catch would be nil, but

---

103 *Report of the Seventh Meeting of Australian, Japanese and New Zealand Scientists on Southern Bluefin Tuna* (15–19 August 1988) (hereinafter 7th Trilateral Scientific Meeting Report), in Trilateral Scientific Meeting Reports Compendium (n. 95) 41, 42–44. This recommendation was criticised by Mr Greig, counsel for Japan, before ITLOS on 19 August 1999 ('How much unnecessary socio-economic damage would have been caused had this plea been heeded?': *Southern Bluefin Tuna Cases* (*New Zealand v. Japan; Australia v. Japan*) (Request for provisional measures) (hereinafter 'Transcript for morning of 19 August') ITLOS doc ITLOS/PV.99/22/Rev.1, 25 <https://www.itlos.org/fileadmin/itlos/documents/cases/case_no_3_4/VRE190899am.corr-rev.2.pdf> accessed 3 August 2015) though he did not mention that the Japanese scientists too supported this recommendation. While the effects of overfishing are usually reversible if catches are sufficiently reduced, instances have been recorded where the stock has failed to recover because the depletion of its numbers has allowed a concomitant increase in the population of another species with which it competes for the same prey, establishing a new equilibrium between the two, e.g. the non-recovery of the sardine population off California: Albert W. Koers, *International Regulation of Marine Fisheries: A Study of Regional Fisheries Organizations* (Fishing News (Books) Ltd 1973) 49, or where the stock is itself preyed upon by some other species—e.g. predation by the grey seal has been advanced as an explanation for the non-recovery of Canadian cod stocks despite the closure of the fishery since 1993, and even though overfishing was the prime cause of the collapse: Caihong Fu, Robert Mohn and L. Paul Fanning, 'Why the Atlantic cod (*Gadus morhua*) stock off eastern Nova Scotia has not recovered' (2001) 58 *Canadian Journal of Fisheries and Aquatic Sciences* 1613, 1622.

104 Caton, McLoughlin and Williams (n. 52) 21, 33.

recommended that if agreement on this drastic action could not be reached, the quotas should nonetheless be reduced by at least half. The management meeting accepted this advice, agreeing on new quotas of 6,250 tonnes for Australia, 8,800 tonnes for Japan and 450 tonnes for New Zealand, the total of 15,500 tonnes being 52 per cent lower than the previous year.[105] Even so, the catch rates of Japanese longliners operating off New South Wales in 1989 were only a third of those recorded in 1980 and 1981.[106] This led the three States to institute a further round of cuts that year, reducing the TAC by 24 per cent to 11,750 tonnes, but while Australia and Japan took respective "national allocations" of 5,265 tonnes and 6,065 tonnes, what New Zealand accepted was a "voluntary catch limit", which it saw as less binding than a national allocation, of 420 tonnes.[107] This can be seen in the resolution passed at their 1990 management meeting in which:

> The three parties confirmed their intentions to set a global quota of 11,750 metric tonnes for 1990/91 with the following national allocations:
> Australia 5,265 tonnes
> Japan 6,065 tonnes
> New Zealand confirmed its intention to limit its own catch to 420 tonnes, noting that Australia and Japan had expressed the view that in the circumstances where the global quota was increased, New Zealand would lift its own catch to 450 tonnes.[108]

In the 1990 trilateral management meeting Japan favoured considering socio-economic factors in setting the TAC, mentioning this in its opening statement:

> There is no single way to promote recovery of SBT stock; we should adopt the method which causes least socio-economic friction. The reality is that extreme measures may look good and be welcomed by the average person but can cause tremendous economic loss to fishery operators.[109]

105 Ibid. 21, 34.

106 Ibid. 23.

107 This distinction, which did not last beyond the entry into force of the 1993 Convention (n. 4) in 1994, is of no significance for present purposes.

108 'SBT Trilateral Management Discussions—Resolution' (ca 1990, unpublished, copy held by author extracted from DPIE files) 1.

109 'Draft transcript of Japanese opening statement, 25 September 1990' (unpublished, copy held by author extracted from DPIE files) 1. Since documents from this source generally have no clear indication that they were ever provided to Japan and New Zealand for comment—or, in the case of statements, that they were delivered in the form written—

Despite the 1989 catch limits having been expressed to be only for one year, agreement was reached to have them do duty again for 1990.[110] As they subsequently remained in place for many years, there is much irony in hindsight in the way the decision to retain them again for 1991 as default limits was qualified:

> Unless there is clear scientific evidence of recovery in the parental biomass, or scientific evidence that required further reductions to be made in quota levels, the global quota and national allocations will remain unaltered for the following year. All parties shared the view that there was a need to avoid being locked into an inappropriate global catch limit.[111]

Although there was encouraging early evidence that the reduced catch limits since the mid-1980s were assisting the cause of the stock's recovery, this was shown to be misleading in 1992 by new information on the growth rates of SBT:

> Under all interpretations of growth, the analyses show that the parental biomass is expected to remain below the 1980 level for many more years, so that the risk of abrupt recruitment decline remains high. Similar to last year, from a biological viewpoint a decrease in catch level is highly desirable.[112]

The next best course of action would be to refrain from increasing the level of catch

> until such time as the parental biomass returns to at least the 1980 level, unless this is part of an agreed stock rebuilding strategy that can be shown to have a high probability of returning the stock to biologically safe levels.[113]

---

they should be treated as no more than an Australian written record of the proceedings with only commensurately qualified confidence to be placed in their accuracy.

110 'Southern Bluefin Tuna Quotas Set' (Media Release PIE 89/328K, John Kerin MP, Minister for Primary Industries and Energy, 24 November 1989) 2.

111 'SBT Trilateral Management Discussions—Resolution' (n. 108) 1.

112 *Report of the Twelfth Meeting of Australian, Japanese and New Zealand Scientists on Southern Bluefin Tuna: Report to Management* (13–19 October 1993) (hereinafter 12th Trilateral Scientific Meeting Report; Attachment C to *Southern Bluefin Tuna Trilateral Management Discussions (Conclusion)* (October–November 1993) (unpublished, copy held by author extracted from DPIE files) (hereinafter 1993 Conclusion)) 16 [16].

113 Ibid. 17–18 [22].

Thus for the remainder of the period preceding the entry into force of the 1993 treaty establishing the CCSBT (hereunder referred to as the 1993 Convention),[114] the trilateral catch limits were left at the level agreed in 1989, and in 1994 at the first meeting of the new commission it confirmed the 1989 TAC and national allocations.[115] The 1993 Convention by Article 8, paragraph 3 directs the CCSBT, "[f]or the conservation, management and optimum utilisation" of SBT, to "decide upon a total allowable catch and its allocation among the Parties" unless it decides upon "other appropriate measures"; it may also "if necessary, decide upon other additional measures". Paragraph 6 of the same article directs that, in doing so, it must "take full account" of the report and recommendations of the Scientific Committee established under Article 9, while paragraph 7 mandates that "All measures decided upon under paragraph 3 shall be binding on the Parties."

8 Gestation of the Dispute—Interpretation of CPUE

Although formally dating from 31 August 1998, when Australia and New Zealand invoked Article 16 of the 1993 Convention,[116] the proximate cause of the dispute, Japan's unilateral experimental fishing programme, had a significant pre-history and an underlying chain of causation of its own. This included both scientific disagreement and fundamental differences in fisheries management philosophy between Australia and New Zealand on the one hand and Japan on the other.

---

114 (n. 4). The 1993 Convention was negotiated by Australia, Japan and New Zealand from 1988 to 1992 at six meetings of the working group established by the trilateral meeting of 1987 for this purpose. Instruments of ratification were deposited in 1994 for Japan, New Zealand and Australia on 8 April, 9 May and 20 May respectively (see the website maintained by the Australian Government as depositary of the Convention, <http://www.austlii.edu.au/au/other/dfat/treaty_list/depository/sbtuna.html> accessed 5 August 2015), and by art. 17(2) it entered into force on the last of these dates.

115 CCSBT, *First Meeting of the Commission for the Conservation of Southern Bluefin Tuna (Conclusion)* (May 1994) (unpublished, cited in A.L. Serdy, *Rights and Obligations of New Entrants into the Southern Bluefin Tuna and Other International Fisheries* (Australian National University PhD thesis, 2008, 91n <https://digitalcollections.anu.edu.au/bitstream/1885/9032/3/02Whole_Serdy.pdf> accessed 7 April 2016); hereinafter CCSBT1 Report) 2.

116 *Southern Bluefin Tuna Cases* (*Australia and New Zealand v. Japan*) (Award on Jurisdiction and Admissibility, 2000) XXIII RIAA 1 (hereinafter Annex VII Tribunal Award) 4 [1].

The scientific disagreement which was to mark the next decade became apparent in 1989 when Australian and Japanese scientists presented projections of future parental biomass assuming the indefinite continuation of the then current quotas. All Japanese projections depicted rising trends while the Australian projections were mixed, depending on whether recruitment after 1981 was assumed to return to the average level for the assumed stock-recruitment relationship or remain at more recent low estimates. As a consequence, there was no agreement on future parental biomass and recruitment trends from the projections. The Australian and New Zealand scientists took the view that, once other indicators from the fishery were taken into account, recovery of the SBT stock was less probable than its continued decline. Their Japanese counterparts alleged that the Australian projections were deliberately pessimistic and unreasonable. The Australian scientists rejoined that the projected recovery depended on optimistic assumptions about recruitment,[117] but did not deny the scope for much larger catch from the SBT stock if the parental biomass could be rebuilt to a sufficiently high level.[118] By 1990 the

117 Caton, McLoughlin and Williams (n. 52) 27; see also the more detailed defence of Australian assumptions and critique of Japanese ones at 29.

118 The necessary implication of this is that the parental biomass was well below the level that would produce the maximum sustainable yield (MSY): see Figure 1 in the Appendix. Although Australia said at the CCSBT's 1996 annual meeting (CCSBT, *Report of the Third Annual Meeting (Revised)* (24–28 September 1996) 19 <http://www.ccsbt.org/userfiles/file/docs_english/meetings/meeting_reports/ccsbt_03/report_of_ccsbt3_part1.pdf> accessed 5 August 2015, hereinafter CCSBT3(1) Report) that the 1980 level "corresponds to commonly used thresholds for biologically safe parental biomass", it is described in Tom Polacheck and others, 'An initial variation of management strategies for the southern bluefin tuna fishery' (1999) 56 *ICES Journal of Marine Science* 811, 824 as "a minimum level for rebuilding and not the target level around which the stock is to be maintained". It can hence be concluded that the MSY is some way below the 1981 catch of 45,000 tonnes, though if the proportion of young fish in the catch were reduced, the total could move closer to that figure. This is corroborated by two studies cited by Harry Campbell, Samuel F. Herrick Jr and Dale Squires, 'The Role of Research in Fisheries Management: The Conservation of Dolphins in the Eastern Tropical Pacific and the Exploitation of Southern Bluefin Tuna in the Southern Ocean' (2000) 31 *Ocean Development & International Law* 347, 360, one of which found that global catch of 32,000 tonnes could be sustained by a population in equilibrium at its 1980 level, while the other described notional Australian and Japanese catches of 11,000 tonnes and 28,000 tonnes respectively as having equal impact on the parental biomass and consistent with maintaining it at an "assumed safe level". In 2014 the Extended Scientific Committee of the CCSBT estimated the MSY at between 30,000 and 36,000 tonnes: CCSBT-ESC Report 2015 (n. 70) Table 5. By a 2001 resolution the CCSBT created an Extended Commission and Extended Scientific Committee

Australian scientists were warning that, assuming the then current trends continued, SBT could be described as commercially threatened and might be facing commercial extinction.[119]

Over the 1990s this divergence of scientific views between Australia and New Zealand on one hand and Japan on the other became entrenched. It related much less to the contemporary state of the stock than to its prospects for recovery in the short to medium term—which became crucial once the CCSBT, not long after it was created, adopted a management goal of rebuilding the parental biomass to its 1980 level by 2020.[120] This state of affairs was the main reason why no consensus to alter the quotas fixed in 1989 could be reached among the three States, a pattern developing of Japanese proposals for increases which Australia and New Zealand did not accept. The concomitant frustrations on both sides caused by this impasse culminated in the Japanese experimental fishing programmes of 1998 and 1999 and the related dispute which lasted until 2001, discussed below.

---

in order to accommodate Taiwan: Resolution to Establish an Extended Commission and an Extended Scientific Committee (Attachment I to CCSBT, *Report of the Seventh Annual Meeting* (18–21 April 2001) (hereinafter CCSBT7 Report), <http://www.ccsbt.org/userfiles/file/docs_english/meetings/meeting_reports/ccsbt_07/report_of_ccsbt7.pdf> accessed 6 August 2015.

119 Caton, McLoughlin and Williams (n. 52) 19. Commercial extinction of a fish species denotes not that it has been rendered extinct by commercial fishing pressure, but that its numbers have fallen to too low a level to make it worth targeting, though individuals of the species may still occasionally be taken as bycatch in other fisheries.

120 CCSBT, *Report of the Second Special Meeting* (29 April–3 May 1996) (hereinafter CCSBTSM2 Report) 1 <https://www.ccsbt.org/sites/ccsbt.org/files/userfiles/file/docs_english/meetings/meeting_reports/ccsbt_02/report_of_special_meeting2.pdf> accessed 31 March 2016. In 1992 the three States had set the 1980 level of parental biomass as a reference level for the rebuilding of the stock "as soon as feasible", confirming this in 1993: *Report of the Eleventh Meeting of Australian, Japanese and New Zealand Scientists on Southern Bluefin Tuna* (5–10 October 1992) (hereinafter 11th Trilateral Scientific Meeting Report; Annex 2 to *Southern Bluefin Tuna Trilateral Management Discussions* (*Conclusion*) (October 1992) (unpublished, copy held by author extracted from DPIE files)) 2; 1993 Conclusion (n. 113) 2. At the 1993 management meeting New Zealand had argued that no increase in catch should be permitted until the 1980 parental biomass was again reached, Australia agreeing that there ought to be no increase in catch levels, for scientific purposes or not, until that goal was achieved: *Southern Bluefin Tuna Trilateral Management Discussions—Second Session—Draft Summary Record* (22–26 November 1993) (unpublished, copy held by author extracted from DPIE files) 4.

The fishery indicators for SBT had historically relied heavily on the interpretation of CPUE.[121] While the scientific analyses of the early 1990s were at one in concluding that the SBT parental stock was substantially depleted from its virgin biomass,[122] there were significant discrepancies in their projections for the stock's recovery, Japan's scientists interpreting the CPUE data in an optimistic fashion and those of Australia and New Zealand pessimistically.[123]

Australia and New Zealand were not opposed in principle, however, to experimental fishing. Indeed, in the trilateral period they had supported a very similar concept, when, citing concerns about "the problem of overfishing, other uncertainties, and the concerns at the historically low levels of parental biomass," 1,000 tonnes of Australian quota was "frozen" in 1990–91, of which 300 tonnes were to be used in a scientific research programme to be formulated jointly by the three States.[124] The detail was set out in an annex to their 1990 resolution:

> ... The basic conditions under which the program would operate are as follows:
>
> 1. It will be carried out by Japanese fishing vessels unless it is specifically decided that Australian or New Zealand vessels should be involved.
>
> ...
>
> 5. Japan guarantees that the program will be implemented within the limit of 300 tonnes.

---

121 For an account of other aspects of fisheries science relevant to SBT see generally the Appendix infra.

122 See e.g. the oral submissions to ITLOS of Counsel for Australia, Professor Crawford, Transcript for afternoon of 18 August (n. 73), 34.

123 This began as early as 1989, when Australia noted in the management meeting that catches by other parties had not been included in the virtual population analysis projections (explained infra Appendix n. 296) considered by the scientists, which added to the need for a cautious approach: *Southern Bluefin Tuna Trilateral Management Discussions Eighth Meeting (Summary Record)* (18–21 September 1989) (unpublished, copy held by author extracted from DPIE files) 27. When the meeting reconvened a month later Australia said the Taiwanese and Korean catch figures provided by Japan were substantially lower than its own estimate; Australia would thus not accept without further discussion the statement by Japan that future parental biomass estimates still increased even with catches by Taiwan and Korea included in the estimates: *Draft Summary Record Reconvened Trilateral Management Meeting for SBT* (8–18 October 1989) (unpublished, copy held by author extracted from DPIE files) 3–4 (hereinafter October 1989 Draft Summary Record).

124 'SBT Trilateral Management Discussions—Resolution' (n. 108) 1.

6. Japan will bear all reasonable costs associated with the research program....
7. If the understandings in relation to the program's implementation are not met, then the allocation of Australian quota can be cancelled by Australia immediately....[125]

Nor is experimental fishing of a highly depleted stock in itself objectionable; when commercial catches have been sharply restricted, experimental catch may be required in order to be able to monitor biomass trends. Where, however, the experiments involve substantial additional catch, managers must balance their information needs for improved (and less risky) future decision-making against the short-term increased biological risks associated with increased catches.[126] At the 1993 management meeting Japan proposed the use of non-commercial quota for a programme to monitor the fishery in real time, but New Zealand noted that this had been discussed and rejected at previous meetings. Given the increasingly serious concerns about the status of the stock, Australia and New Zealand said they could not consider an allocation of scientific quota outside the trilateral commercial catch limits.[127]

The first hint of experimental fishing along the lines ultimately pursued by Japan came in a suggestion by an invited expert at the 1994 trilateral scientific meeting that difficulties in the interpretation of CPUE could be addressed by the division of the TAC between currently fished 5°-square areas and areas where no fishing had occurred in recent years.[128] These difficulties had arisen because

> [t]he Japanese long-line fishery has contracted in recent years, and substantially in 1993 in areas fished each month. Such a contraction results in no CPUE data from many area-months and, because no information is available from unfished areas, increased uncertainty about total stock abundance. This uncertainty is largely responsible for the wide range of

125 Ibid. 5 (Annex 2, 'SBT Special Research Program').

126 Polacheck (n. 3) 283–284.

127 *Southern Bluefin Tuna Trilateral Management Discussions—First Session—Draft Summary Record* (21–23 October 1993) (unpublished, copy held by author extracted from DPIE files) 5.

128 Polacheck (n. 3) 287.

interpretations of recent CPUE increases of juveniles within the fished area-months.[129]

Two of the CPUE series developed for SBT represented bounds for the relative density in unfished areas in a given year. The lower bound, known as the variable squares hypothesis, took the density in the unfished squares as zero, in other words it assumed no SBT at all were present. The hypothesis was intended as a simplified representation of the effect of fishing effort being concentrated in areas of highest density (that is, it assumed that fishing masters have perfect knowledge of where the fish are).[130] The upper bound, referred to as the constant squares hypothesis, equated the relative density in unfished squares with that in the areas fished, reflecting an underlying assumption that fishing masters had no such knowledge at all and instead fished at random.[131]

129 13th Trilateral Scientific Meeting Report (n. 54), 3 [12]. The Report of the previous year's scientific meeting had noted a recent pattern of longline fishing effort moving from areas that historically provided high catches of large and valuable fish to areas where smaller fish were now being caught: 12th Trilateral Scientific Meeting Report (n. 112) 8 (Appendix 1, 'Status of the stock and fishery indicators', [4]).

130 Over the years, the CCSBT Scientific Committee also considered a number of other hypotheses. Of these the one given the most weight in stock assessments is based on a geostatistical approach which estimates the density in areas without data as a function of the relative density in neighbouring areas and time periods: Polacheck (n. 3) 287. Note that, if the aim were purely to model fishing effort, the assumption of zero density would not appear to be justified. Even if a vessel could move instantly and costlessly from one area to another, the logical consequence of the assumption that fishing masters are systematically able to target areas of higher density is simply that no unfished square should have a density higher than the lowest density of any fished square. The justification for zero density, if there is one, is that zero is a peremptory lower bound that is useful for modelling purposes. That is, if models that give a certain non-zero weight to the premise that there are no fish in unfished squares are shown to be consistent with the observed data, as was the case here, even the fact that the premise is accepted or proved to be wrong should not, as a proposition of science let alone of law, prevent their use. Not associated with the applicants, D.S. Butterworth, J.N. Ianelli and R. Hilborn, 'A Statistical Model for Stock Assessment of Southern Bluefin Tuna with Temporal Changes in Selectivity' (2003) 25 *African Journal of Marine Science* 331, 354 accept that the intent of the variable squares hypothesis was to provide a lower bound rather than a realistic scenario.

131 See also Maguire (n. 3) 204–205 and Douglas S. Butterworth and Andrew J. Penney, 'Allocation in High Seas Fisheries: Avoiding Meltdown' in A.I.L. Payne, C.M. O'Brien and S.I. Rogers (eds), *Management of Shared Fish Stocks* (Blackwell Publishing Ltd 2003) 165, 178 and Japan's written submission to ITLOS, *Southern Bluefin Tuna (Australia and New Zealand v. Japan) Response and Counter-Request for Provisional Measures submitted by Japan* ITLOS doc F4/155/198 <https://www.itlos.org/fileadmin/itlos/documents/

A month later Japan proposed at the First Meeting of the CCSBT to instruct the Scientific Committee to formulate a real-time monitoring programme for the 1995 season to cover areas and periods outside the commercial fisheries with a minimum level of catch required for scientifically valid research, the catch taken in the course of the programme not to be counted against the global quota.[132]

At the following Scientific Committee meeting, Japanese scientists tabled a three-page proposal for extensive experimental fishing, which Polacheck, at the time a member of Australia's scientific team, criticises for lacking details on the experimental design, its analysis, the way the results would be used to manage the stock and the basis for the magnitude of the catch sought.[133] At an April 1995 informal meeting Japan elaborated on its research quota proposal, arguing that it should be allocated to vessels to fish in areas and seasons not currently fished, so that data would be available for resolving uncertainties in the stock assessment, such as the "constant vs variable squares" question. Australia and New Zealand replied that it was unacceptable to contemplate quota increases, whether for commercial or research purposes, until a clear management strategy incorporating reference points had been established.[134]

If Australia and New Zealand did not at first take the concept of an experimental fishing programme seriously, the explanation for this lies in the context in which it was introduced by Japan into the Commission proper. At the Second Meeting of the CCSBT in September 1995, while Australia and New Zealand took the position that the TAC should be left as it was, with no change to Members' national allocations, Japan sought an increase in the TAC of 6,000 tonnes without mentioning its earlier research quota proposal.[135] As no consensus could be reached on the TAC and national allocations, the CCSBT decided to continue its deliberations on this matter the following month by convening a special meeting, as provided under Article 6, paragraph 5 of the

cases/case_no_3_4/statement_response_japan_eng.pdf> accessed 14 August 2015 (hereinafter Japanese written statement to ITLOS), for explanations of the constant or variable squares problem.

132 CCSBT1 Report (n. 115) 1 and Annex 1, Part 2, [1], [3].

133 Polacheck (n. 3) 288.

134 CCSBT1 Report (n. 115) 8.

135 CCSBT, *Report of the Second Annual Meeting* (12–15 September 1995) 1 <http://www.ccsbt.org/userfiles/file/docs_english/meetings/meeting_reports/ccsbt_02/report_of_ccsbt2.pdf> accessed 6 August 2015. Japan did not specify how this increase should be divided among the Members' national allocations.

1993 Convention.[136] At the Special Meeting, Japan renewed its call for a 6,000-tonne increase in the TAC, but now offered alternatives: a 6,000-tonne experimental fishing quota above the TAC to be fished annually for three years, or a combination of additional TAC and experimental quota totalling 6,000 tonnes *per annum*.[137]

With the impasse on setting a TAC for the 1996 season persisting, at a subsequent session of this meeting in January 1996 it was agreed that work to evaluate the possible implementation of an experimental fishing programme was warranted.[138] In a further special meeting in 1996, the CCSBT agreed on a process both to evaluate the impact of additional removals and for designing and evaluating proposals for experimental fishing.[139] The relevant attachment set out the following agreed matters:

> [Experimental fishing] could happen where there is agreement within the Commission that the risks of such extra removals are outweighed by the benefits. [...] Prior to the Commission deciding to proceed with any experimental fishing program it will need to agree on the way in which results coming from the program would be incorporated into the stock assessment and the future management decision-making for the fishery.[140]

In addition, the development, evaluation, implementation and analysis of the programme should be collaborative and agreed among all parties; prior to any such programme being implemented, agreement should be reached on the specific criteria for determining whether any additional removals would jeopardise stock recovery. The criteria for judging an experimental

---

136 1993 Convention art. 6(5) provides that "Special meetings of the Commission shall be convened by the Chair at the request of a Party supported by at least two other Parties." A special meeting may consider any matter relevant to the 1993 Convention: art. 6(6).

137 CCSBT, *Report of the First Special Meeting* (3–6 October 1995) 1 <http://www.ccsbt.org/userfiles/file/docs_english/meetings/meeting_reports/ccsbt_02/report_of_special_meeting1_part1.pdf> accessed 6 August 2015, and the proposal appended to Annex 2 ('<<Japan's Proposal>> on Special Experimentai [*sic*] Fishing Arrangements (Additional Measures) for the Special Meeting of CCSBT October 1995 Canberra').

138 CCSBT, *Report of the Reconvened First Special Meeting* (17–19 January 1996) 2 <http://www.ccsbt.org/userfiles/file/docs_english/meetings/meeting_reports/ccsbt_02/report_of_special_meeting1_part2.pdf> accessed 6 August 2015.

139 'Objectives and principles for the design and implementation of an experimental fishing program' (Attachment C to CCSBTSM2 Report (n. 120)).

140 Ibid. 1.

fishing programme "should be derived from management objectives" and any such programme "should be designed to deliver scientifically valid and meaningful results", with "appropriate monitoring of any program, designed and conducted in a collaborative manner amongst the parties."[141]

The evaluation process was specified in some detail, entailing three steps. Step 1 envisaged the reaching of agreement on the range of uncertainty to be considered in evaluation of experimental fishing proposals, and on the weight to be placed on the various options within that range. In Step 2 there would be an evaluation of the effect of changed catch levels on the chance of recovery, including the effects on recovery of resolving the main uncertainties, and suggestions as to how such resolution might be achieved. The aim of Step 3 would be to determine the most effective use of experimental catch, including the type of information provided by experimental fishing and how that information would be used by managers. This would involve empirical analysis of past fishing experience on possible results from the proposed experiment, simulation of those results and their analysis by agreed methods.[142]

9 Impasse in the CCSBT

Despite holding numerous workshops and meetings from 1996 to 1998 to develop and evaluate experimental fishing proposals based on the agreed objectives and principles and the three-step evaluation process, the three States were unable to reach consensus on whether or how such fishing should proceed. A series of proposals for experimental fishing put forward by Japan were not accepted by Australia or New Zealand, and thus by the CCSBT itself, because they were viewed as not meeting the objectives and principles agreed earlier in 1996. They considered that the proposed experimental fishing was misdirected relative to the true impacts of uncertainty within the stock assessment and that the experimental design and analysis were fundamentally flawed. This was because the experiment targeted only one source of uncertainty which, even if it were resolved, would only slightly narrow the differences among the parties about stock recovery and appropriate TAC levels. Thus, in the applicants' view, the objectives of the experiment did not justify the significant increased risk to the stock, especially when other mechanisms for reducing uncertainty were available that would require little or no addi-

141 Ibid. 2–3.

142 'Scientists report for evaluating the impact of additional removals for experimental fishing on the recovery of the SBT stock' (Attachment D to CCSBTSM2 Report (n. 120)) 1–2.

tional catch.[143] The failure of this process is attributed by Polacheck in part to the

> very different attitudes about the importance and rigour with which the criteria and process needed to be adhered to, as well as ... a difference in perspective about whether the outcome of the process was to provide the basis for deciding on whether an EFP was worth conducting or whether fundamentally, this was already clear and only a technical design was required.[144]

According to Polacheck, the difference in perspective between Australia and New Zealand on one hand and Japan on the other as to what was required in terms of the linkage between the results from an EFP and an agreed management response was a key obstacle to agreement.[145] From the outset, the support of Australia and New Zealand for any such programme was dependent upon a direct linkage: "Australia and New Zealand indicated that they would support collaborative and timely work on designing and evaluating an EFP, linked by means of predetermined management responses to clear management objectives."[146]

Japan's earliest proposals contained no reference to decision rules (pre-agreed catch limit adjustments flowing and calculable directly from the data that would be obtained) or management responses. Though some later ones included a section on decision rules, these did not extend to specific predetermined management responses, and allowed only for increases in TAC if the results were positive, but not for decreases in the TAC, beyond a possible payback of EFP catches, if the results were negative.[147]

143 Statement of Claim (n. 5) [17].

144 Polacheck (n. 3) 288. EFP as used here and below stands for "experimental fishing program[me]".

145 Ibid. 291, where a number of other factors are listed on which differences in perspective prevented agreement from being reached.

146 CCSBTSM2 Report (n. 120) 1.

147 Polacheck (n. 3) 290–291. Japan argued that, over a timeframe longer than that of the additional short-term catch it proposed, the added risk of not meeting the recovery target could be negated by a compensating decrease in catch. It thus offered to make such a compensatory reduction if its experimental fishing catches were shown to be detrimental to the stock. Australia and New Zealand found this unsatisfactory because Japan had put forward no standard or method for measuring detriment, and placed the onus of proof on them to show that it had occurred. Since the effects would not manifest themselves in the short term, without settling fixed time horizons and decision criteria, the default

While the work on the EFP was underway, Japan continued to insist on a substantial increase in the TAC or alternatively a large experimental fishing quota. Australia and New Zealand were not able to accept this before the work on the EFP was completed, with the result that under its consensus rule for decision-making,[148] the CCSBT became unable to set a TAC and national allocations: Japan was able to prevent the adoption of any TAC that did not meet its demand for an increase, and Australia and New Zealand were able to prevent the adoption of an increased TAC. In 1996 and 1997, for the sake of continued access to the AFZ for its fleet, Japan relented just in time to secure this,[149] but from 1998 this enticement was no longer sufficient and it was not until 2003 that a TAC could again be decided when, principally as a result of a glut of tuna on its market,[150] Japan abandoned its demand. It was Japan's 1998

outcome would always be to prolong the period of additional catches and delay initiation of the compensating reductions. While the marginal increase in detriment of waiting one more year would always tend to be small, the cumulative effect of doing so repeatedly could be devastating. A further weakness of Japan's proposal was that the risks at low stock sizes were not symmetrical nor well estimated, so that decrease of future catch might be of no use if the stock collapsed in the meantime: see infra Appendix, text following n. 292 and Polacheck (n. 3) 292.

148 1993 Convention art. 7 (n. 114) provides that "Each Party shall have one vote in the Commission. Decisions of the Commission shall be taken by a unanimous vote of the Parties present at the Commission meeting." In other words, each Member is able to veto any decision on TAC and national allocations not to its liking, provided only that it attends the meeting at which the decision is made.

149 Before the Annex VII Arbitral Tribunal Japan attributed its reluctant agreement to leave the TAC and national allocations unchanged in the years leading up to the dispute to Australia's longstanding policy of allowing access to its ports for Japanese fishing vessels only while their catch of SBT was subject to a negotiated limit: see its memorial for the preliminary objections phase of the dispute, Government of Japan, *Cases Concerning the Convention for the Conservation of Southern Bluefin Tuna* (*Memorial on Jurisdiction of Japan*) (hereinafter Japanese memorial) 23 [53] <https://icsid.worldbank.org/apps/ICSIDWEB/Documents/Memorial%20on%20Jurisdiction%20of%20Japan.pdf> accessed 5 August 2015), and the oral submissions of its counsel, Professor Ando, *Southern Bluefin Tuna Case* (*Australia and New Zealand v. Japan*) (*First Round Presentation of Japan, May 7, 2000*) (hereinafter Transcript for 7 May) 34–35 <https://icsid.worldbank.org/apps/ICSIDWEB/Documents/First%20Round%20Presentation%20of%20Japan_May%207_2000.pdf> accessed 5 August 2015.

150 Japan reported a severe decline in the price of tuna, caused by the combined effect of the continued high level of imports of these species and a fall in demand for them brought about by Japan's long economic recession: 'Review of Southern Bluefin Tuna Fisheries of Japan in the 2003 Fishing Season' (Attachment 8–3 to CCSBT, *Report of the Extended Commission of the Tenth Annual Meeting of the Commission* (7–10 October 2003) (herein-

decision to fish experimentally on its own outside this framework that provoked the dispute under UNCLOS and the 1993 Convention.

10 Japan's Experimental Fishing

In February 1998, Japan indicated that it would fish commercially at the level of national allocation fixed for its most recent fishing year (March 1997 to February 1998) and that it would take an additional 2,010 tonnes of SBT for three years, for the purposes of unilateral experimental fishing. Between March and June 1998, talks were held among Australia, Japan and New Zealand, but these failed to resolve the differences over the Japanese proposal. On 1 June 1998, Japan made a revised proposal for "pilot" experimental fishing of 1,400 tonnes to start on 1 July that year, to precede its three-year programme. Australia and New Zealand replied that the proposal was unacceptable by reference to the agreed criteria, and requested Japan not to proceed. Despite those requests, Japan undertook its pilot programme in the southern Indian Ocean from 10 July to 31 August 1998, resulting in a combined commercial and experimental catch of SBT 1,464 tonnes in excess of its previously agreed national allocation.[151]

On 31 August 1998 Australia and New Zealand invoked the dispute settlement procedures under the 1993 Convention, requesting urgent consultations and negotiations under Article 16, paragraph 1. This led to consultations in December 1998, but no solution was reached beyond an agreement to continue intensive consultations in 1999 with a view to finding an acceptable joint programme of experimental fishing for that year.[152] Again, however, no agreement was reached between the parties in this framework.[153] Although the

after CCSBT-EC Report 2003; Appendix 3 to CCSBT, *Report of the Tenth Annual Meeting of the Commission* (7–10 October 2003))), 10 [51] <http://www.ccsbt.org/userfiles/file/docs_english/meetings/meeting_reports/ccsbt_10/report_of_ccsbt10.pdf> accessed 3 August 2015. See further infra, text at (nn. 235–236).

151 Statement of Claim (n. 5) [13].

152 For this purpose the parties established an Experimental Fishing Program Working Group of managers and other officials, national scientists and industry representatives as well as three invited external scientists. The Working Group's terms of reference required that account be taken of the 1996 Objectives and Principles and listed eight specific tasks that were similar and consistent with the basic development and evaluation steps agreed on previously: Polacheck (n. 3) 288.

153 Polacheck, ibid. 289 found this unsurprising given the complexity of the tasks that were assigned to the Working Group and the short timeframe of four meetings between February and April 1999 (see CCSBT, 'List of Past CCSBT Meetings' <http://www.ccsbt

Working Group's terms of reference required it to "decide on appropriate decision rules governing the interpretation of the results for the management and conservation of SBT", this proved impossible because the CCSBT had not developed a management decision-making framework.[154]

Japan informed Australia and New Zealand in May 1999 that it planned to reinstitute its experimental fishing programme; Australia and New Zealand responded on 1 June that its renewal would be considered a termination of the negotiations. Japan proceeded with its experimental fishing programme, but simultaneously maintained that it did not intend to curtail negotiations.[155] On 23 June 1999, Japan requested that the dispute with Australia be resolved by mediation pursuant to the 1993 Convention, and made the same request of New Zealand the following day. Both applicants replied on 30 June 1999 that they were willing to submit the dispute to mediation if Japan discontinued its experimental fishing programme.[156] When Japan rejected this, Australia and New Zealand notified Japan of their decision to commence compulsory dispute resolution under Part XV of UNCLOS, submitting the dispute to an arbitral tribunal constituted under Annex VII, the default dispute settlement mechanism, instituting these proceedings by almost identical statements of claim on 15 July 1999.[157] Pending the constitution of such a tribunal, they made similarly parallel requests to ITLOS to prescribe provisional measures, as provided by Article 290, paragraph 5 of UNCLOS.[158] The provisional measures requested

---

.org/userfiles/file/docs_english/meetings/ccsbt_previous_meetings.pdf> accessed 6 August 2015).

154 The terms of reference are cited in Polacheck, ibid. 291. Instead, the Working Group proposed a "metarule", by which the experiment could continue into 1999 but not proceed into future years unless decision rules were first agreed; Australia and New Zealand found this unacceptable: Japanese memorial (n. 149) 39 [81]; *Southern Bluefin Tuna Cases* (*Australia and New Zealand v. Japan*) (Reply on Jurisdiction [of] Australia and New Zealand) 98 [A33] <https://icsid.worldbank.org/apps/ICSIDWEB/Documents/Reply%20on%20Jurisdiction%20of%20Australia%20and%20New%20Zealand.pdf> accessed 6 August 2015.

155 Annex VII Tribunal Award (n. 116) 13 [26].

156 Japanese memorial (n. 149) 44–45 [90], 46 [94].

157 Under UNCLOS art. 287 a State may at any time declare that it accepts one or more of four means of compulsory dispute settlement: ITLOS, the ICJ, an arbitral tribunal established under Annex VII or, for certain categories of disputes including fisheries disputes, a special arbitral tribunal established under Annex VIII. If a State has not selected a particular means of dispute settlement, as at the time of institution of proceedings in this case was true of all the parties to the dispute, it is deemed to have accepted arbitration under Annex VII.

158 UNCLOS art. 290 (5) provides that, pending the establishment of an arbitral tribunal under Annexes VII or VIII, ITLOS, if it considers that *prima facie* the arbitral tribunal

included the cessation of Japan's experimental fishing programme, the restriction of Japan's future catches to the last agreed level, less the experimental catch for 1998 and 1999, and other orders protecting the rights of the parties.[159]

11 The Relevant Law Identified in the Statements of Claim

Despite having instituted the dispute under the dispute settlement provision (Article 16) of the 1993 Convention, the Applicants did not cite any specific provision of that treaty; rather, they claimed that Japan's experimental fishing was in breach of Articles 64 and 116–119 of UNCLOS.[160]

The text of Article 64 of UNCLOS:

Article 64
Highly migratory species

> 1. The coastal States and other States whose nationals fish in the region for the highly migratory species listed in Annex I shall co-operate directly or through appropriate international organizations with a view to ensuring conservation and promoting the objective of optimum utilisation of such species throughout the region, both within and beyond the exclusive economic zone. In regions for which no appropriate international organization exists, the coastal States and other States whose nationals harvest these species in the region shall co-operate to establish such an organization and participate in its work.
> 2. The provisions of paragraph 1 apply in addition to the other provisions of this Part.

is the result of a compromise between fishing States which wanted to create a special regime for species of fish that undertake extensive transoceanic

would have jurisdiction to hear the case, may prescribe any provisional measures it considers appropriate to preserve the respective rights of the parties or to prevent serious harm to the marine environment.

159 *Southern Bluefin Tuna Cases* (*New Zealand v. Japan; Australia v. Japan*) (Provisional Measures, Order of 27 August 1999) [1999] ITLOS Rep 280 (hereinafter ITLOS Order), 288–289 ([31] (New Zealand) and [32] (Australia)).

160 Statement of Claim (n. 5) [45] and [69(1)], with fuller reasoning at [53–54] (art. 64), [59–62] (art. 116), [55] (art. 117), [56] (art. 118), [57–58] (art. 119). See also infra, section 16 on specific issues within these provisions that a hearing on the merits might have resolved.

migrations,[161] and coastal States which saw no reason why such species should be treated differently from the others occurring in their EEZs.[162] As the text indicates, it applies also on the high seas, which is where Japan's experimental fishing took place. Article 64, along with the remainder of Part V of UNCLOS, was found "consonant with general international law" by a chamber of the ICJ in the *Gulf of Maine Case*.[163]

Articles 116 to 119 fall within and constitute the bulk of section 2 of Part VII of UNCLOS bearing the heading "Conservation and Management of the Living Resources of the High Seas".[164] These articles must also be read with Article 87, which reads, so far as material:

> 1. The high seas are open to all States, whether coastal or land-locked. Freedom of the high seas is exercised under the conditions laid down by the Convention and by other rules of international law. It comprises, *inter alia*, for both coastal and land-locked States:
>
> …
>
> (e) freedom of fishing, subject to the conditions laid down in section 2.

---

161 According to William T. Burke, *The New International Law of Fisheries: UNCLOS 1982 and Beyond* (Clarendon Press 1994) 199, it was chiefly tunas that Japan and the United States had in mind.

162 Myron H. Nordquist, Satya N. Nandan, and Shabtai Rosenne (eds), *United Nations Convention on the Law of the Sea, 1982: A Commentary*, vol. II (Martinus Nijhoff Publishers 1993) 649–650 (hereinafter *Virginia Commentary*). Because the duty of cooperation extends to the EEZ, it cannot avoid impinging to some degree on the otherwise largely unfettered discretion of the coastal State in fixing the allowable catch of species that occur in its EEZ, though it falls short of requiring access by other States to any surplus that is beyond the coastal State's own capacity to harvest there—see arts 61–62 and the discussion in Burke (n. 161) 213–217 and Francisco Orrego Vicuña, *The Changing International Law of High Seas Fisheries* (Cambridge University Press 1999) 26–28. In practice, coastal States have displayed a consistent tendency to adopt more conservative stances on permissible levels of fishing pressure than distant-water fishing States, which makes it more likely that when disputes over a highly migratory species arise they will be about the exploitation of it on the high seas rather than the intensity of the coastal State's fishing for it in the EEZ—as indeed occurred in the SBT dispute itself.

163 *Case concerning Delimitation of the Maritime Boundary in the Gulf of Maine Area (Canada/United States of America)* (Judgment) [1984] ICJ Rep 246, 294 [94]. R.R. Churchill and A.V. Lowe, *The Law of the Sea* (3rd edn, Manchester University Press 1999) 161–162 conclude that the coastal State's EEZ rights have attained customary status, but that this has not happened for the obligations in arts 61–62.

164 The only other provision in this section, art. 120, is confined to marine mammals and thus of no consequence for SBT.

> 2. These freedoms shall be exercised by all States with reasonable regard to the interests of other States in their exercise of the freedoms of the high seas,. . .

Article 116, headed "Right to fish on the high seas", provides that:

> All States have the right for their nationals to engage in fishing on the high seas subject to:
> (a) their treaty obligations;
> (b) the rights and duties as well as the interests of coastal States provided for, *inter alia*, in article 63, paragraph 2 and articles 64 to 67; and
> (c) the provisions of this section.

Although Article 119, paragraph 1 directing States, in "determining the allowable catch and establishing other conservation measures for the living resources in the high seas" to:

> (a) take measures which are designed, on the best scientific evidence available to the States concerned, to maintain or restore populations of harvested species at levels which can produce the maximum sustainable yield, as qualified by relevant environmental and economic factors, including the special requirements of developing States, and taking into account fishing patterns, the interdependence of stocks and any generally recommended international minimum standards, whether subregional, regional or global;
> (b) . . .

does not state *expressis verbis* that the measures must be jointly decided by all interested States, that is its effect when read in conjunction with Article 118:[165]

> States shall cooperate with each other in the conservation and management of living resources in the areas of the high seas. States whose nationals exploit identical living resources, or different living resources in the same area, shall enter into negotiations with a view to taking the measures necessary for the conservation of the living resources concerned. They shall, as appropriate, cooperate to establish subregional or regional fisheries organizations to this end.

---

165 As urged in *Virginia Commentary* (n. 162) vol. III (Martinus Nijhoff Publishers 1995) 309–310.

Likewise, Article 117:

> All States have the duty to take, or to cooperate with other States in taking, such measures for their respective nationals as may be necessary for the conservation of the living resources of the high seas.

complements the obligation of cooperation set out in Article 64, something which is also true of Article 119, paragraph 2:

> Available scientific information, catch and fishing effort statistics, and other data relevant to the conservation of fish stocks shall be contributed and exchanged on a regular basis through competent international organizations, whether subregional, regional or global, where appropriate and with participation by all States concerned.

While Articles 64 and 117 to 119 have an obvious relevance to the merits of the dispute (considered in section 16 below), it may be noted here that the allegation of breach of Article 116 is puzzling, as this provision appears to create no obligation as such that is capable of being breached, except the implied one of other States, correlative to the fishing State's right, to refrain from prevention of, or interference with, the exercise of the right. Even if Australia and New Zealand were arguing that breaches of the other provisions of UNCLOS mentioned in paragraph (b) or (c) of Article 116 deprived Japan of the right to fish for SBT at all on the high seas,[166] that would entail no more than asking the Annex VII Tribunal to apply Article 116 as a consequence of those breaches, not a request to remedy an independent breach of Article 116. The better view must be that the obligations set out in paragraphs (a) to (c) condition the manner of exercise of the right, not its very existence.

---

166 In all likelihood the applicants were not asserting this. Their Statement of Claim (n. 58) argued only that Japan's conduct in undertaking experimental fishing unilaterally was "not authorised or permitted by Article 116" (the phrase occurs in both [60]—because it was "calculated to defeat the object and purpose of the 1993 Convention" and [62]—because it was "in breach of its obligations under Articles 117, 118 and 119"), and the orders they sought called merely for limitation of Japan's catch of SBT, not its cessation: at [69(2)(c)–(d)]. Nor is it clear where such an argument would end: if a State "breaches" art. 116 in respect of one species, does it thereby lose the right for its nationals to engage in fishing on the high seas for *any* species?

12 The ITLOS Hearing and Order

ITLOS heard the requests for provisional measures on 18–20 August 1999 at its seat in the German port of Hamburg.

Japan objected to the jurisdiction of ITLOS on the ground that the Annex VII Tribunal, once constituted, would itself lack jurisdiction to decide the dispute. It argued that the dispute concerned the interpretation or application not of UNCLOS but of the 1993 Convention, and in the alternative that Australia and New Zealand had not met the preliminary requirements of UNCLOS Article 283 for invocation of the compulsory dispute settlement mechanism.[167] Japan contended *inter alia* that the growing catches of SBT by non-members of CCSBT were reasons in themselves to conduct experimental fishing:

> No state, I submit, will join the 1993 Convention unless it can be given a reasonable quota to make joining viable.
>
> If the EFP were to demonstrate that Applicants' pessimistic projections are realistic it will be practically difficult to induce non-parties to the 1993 Convention to participate. On the other hand, if the EFP leads to more optimistic conclusions, bringing the Republic of Korea and other non-parties into a regional Southern Bluefin Tuna management arrangement should be feasible.[168]

---

167 ITLOS Order (n. 159) 290 [33] and 294 [56].

168 Oral submissions of the Agent for Japan, Mr Togo, Transcript for morning of 19 August (n. 103) 17. This argument comes close to begging the question, resting as it does on the assumption that the experiment would vindicate Japan's optimism, and is undermined by the implication that if the findings were instead pessimistic they would have to be suppressed for policy reasons in order not to make entry unattractive to the prospective new members. This was not the first display of a certain naivety about the consequences of the reduction of scientific uncertainty. At the 1990 trilateral meeting, noting that the recommendations of the scientists as to quota incorporated a discount for uncertainty, Japan's understanding (P. Enright (Australian Fisheries Service), *Draft Summary Record Trilateral Management Meeting for SBT* (25 September–26 October 1990) (unpublished, copy held by author extracted from DPIE files) 10) was that "if the catch is closely monitored and reliable scientific data obtained, so the uncertainties in the models are reduced, then it would be possible to increase catch levels this year." That ignores the possibility that the more reliable information thus obtained might reveal that a lower base level of quota was appropriate, wholly or partly negating the smaller discount than hitherto that could be applied in consequence of the reduced uncertainty—but it is far from clear that Japan grasped this.

ITLOS issued its decision on 27 August 1999. Finding the applicants' arguments persuasive, ITLOS held that it had jurisdiction in the dispute on the basis that the Annex VII Arbitral Tribunal would *prima facie* have jurisdiction over it.[169] Rejecting Japan's argument, ITLOS found that a dispute arising under UNCLOS did exist; the fact that the 1993 Convention might also apply did not preclude recourse to the Part XV procedures.[170] As there was no obligation to persist with negotiations beyond the point where it was clear they would not lead to settlement of the dispute, ITLOS was satisfied that the jurisdictional requirement of Article 283 had been met.[171]

ITLOS admonished the parties that they "should in the circumstances act with prudence and caution to ensure that effective conservation measures are taken to prevent serious harm to the stock of southern bluefin tuna".[172] Although the ITLOS Order did not refer to the precautionary principle, some judges in their Separate Opinions considered it relevant in support of the overall decision.[173]

Pending the Annex VII Tribunal's decision, ITLOS prescribed a number of provisional measures in the *dispositif* of its Order. In particular it ordered that the parties should each ensure that their catches did not exceed the annual national allocations they had last agreed and that they should each refrain from conducting experimental fishing programmes unless the catch was counted against the allocation, except with the agreement of the others. It also

169 ITLOS Order (n. 159) 295 [62].

170 Ibid. 294 [52–55].

171 Ibid. 295 [60–61]. Accord the *MOX Plant Case* (*Ireland v. United Kingdom*) (Provisional Measures, Order of 3 December 2001) (hereinafter MOX Plant Case Order) [2001] ITLOS Rep 95, 107 [60]: "a State Party is not obliged to continue with an exchange of views when it concludes that the possibilities of reaching agreement have been exhausted"), the view of the Annex VII Tribunal infra, text at (n. 197) and *Virginia Commentary* (n. 162) vol. V (Martinus Nijhoff Publishers 1989) 23, which affirms the proposition that a State may submit a case to Part XV procedures whenever it concludes that the current procedures are no longer likely to lead to a settlement, but adds that, if the other party objects, "the tribunal or court to which the matter is submitted will have to decide this preliminary objection to its jurisdiction."

172 ITLOS Order (n. 159) 296 [77].

173 *Southern Bluefin Tuna Cases* (*New Zealand v. Japan; Australia v. Japan*) (Order of 27 August 1999), Sep. Op. Laing [1999] ITLOS Rep 305, 309–315 [12–21]; Sep. Op. Treves [1999] ITLOS Rep 316, 317–319 [8–11]; Sep. Op. Shearer [1999] ITLOS Rep 320, 326–327. Maguire (n. 3) 217 referred to it as a possible manifestation of the fishing industry's "worst fear with the implementation of the precautionary approach".

ordered the parties to resume negotiations without delay with a view to reaching agreement on measures for the conservation and management of SBT.[174]

Since the experimental catch in 1999 was 2,198 tonnes,[175] following the ITLOS Order Japan announced that, as a first instalment towards compliance with subparagraph (d) of the Order, it would reduce its commercial catch for the 1999 season—which was already underway—by 700 tonnes, with the remaining 1,498 tonnes held over to the following season. In the event, its commercial catch fell 11 tonnes short of the revised limit, leaving 1,487 tonnes to be repaid in 2000.[176]

The three parties held negotiations in Canberra in September 1999. As they subsequently reported to ITLOS, progress was made on a number of issues including the introduction of a trade information system to assist in the management of the global catch of SBT and an action plan aimed at bringing the other major exploiters of SBT within the 1993 Convention, but the dispute was not resolved.[177]

13 The Parties' Arguments before the Annex VII Arbitral Tribunal

In parallel with further progress made in meetings of the CCSBT in November 1999 and March 2000, the constitution of the Annex VII Tribunal proceeded pursuant to Article 3 of that annex. It comprised Sir Kenneth Keith (then a judge of the New Zealand Court of Appeal)[178] and Professors Chusei Yamada (at that time the Japanese member of the International Law Commission),[179] Florentino Feliciano (then of the WTO Appellate Body), Stephen Schwebel (the recently retired former President of the ICJ), and Per Tresselt of Norway (then a Judge of the EFTA Court).[180] Japan having indicated that it had objections to both the tribunal's jurisdiction and the admissibility of the applicants' claims,

174 ITLOS Order (n. 159) 297–300 [90.1]. The orders in relation to catch limits and counting towards them of experimental catch and resumption of negotiations are at subparagraphs (c), (d) and (e) respectively. Among the ancillary orders was the submission of an initial report on compliance with the substantive orders by 6 October 1999: 300 [90.2].

175 *Reply on Jurisdiction [of] Australia and New Zealand* (n. 154) 95 (Table 1) and 99 [A37].

176 'Review of Southern Bluefin Tuna Fisheries of Japan in 1999 Fishing Season' (Attachment F-2 to CCSBT7 Report (n. 118)).

177 ITLOS, *Yearbook 1999 Volume 3* (Kluwer Law International 2001) 51.

178 Sir Kenneth was nominated by the applicants under UNCLOS Annex VII art. 3 (b).

179 Professor Yamada was nominated by the respondent under UNCLOS Annex VII art. 3 (c).

180 The other three members were jointly nominated by the parties under UNCLOS Annex VII art. 3 (d); Professor Schwebel was chosen by the parties as president.

a hearing on these preliminary points took place in Washington in May 2000, preceded by the filing of the memorials of both sides.[181]

Despite ITLOS having held that the Annex VII Tribunal would *prima facie* have jurisdiction, Japan continued to argue that its dispute with Australia and New Zealand related exclusively to the 1993 Convention and not UNCLOS, as Article 279 of the latter required.[182] Alternatively, Article 281, paragraph 1 had not been satisfied since the parties had not exhausted their dispute resolution efforts under the 1993 Convention; nor had Australia and New Zealand complied with Article 283, by failing to exchange views with Japan either about the dispute or on termination of negotiations.[183] Australia and New Zealand denied both that the dispute arose solely under the 1993 Convention and that UNCLOS was inapplicable merely because the 1993 Convention was applicable; since both Conventions were relevant, recourse to the UNCLOS Part XV procedures was not precluded.[184] Rather, the applicability of a treaty was a question for "objective judicial or arbitral processes" to determine.[185] They explained that negotiations "over the best part of a year" had been "extensive" and "intensive" and that Japan's continuation of the experimental fishing programme into a second year "was rightly regarded as tantamount to a termination of negotiations", exhausting any possibility of settlement.[186]

Apart from its ultimately successful argument on the relationship between the dispute settlement provisions of UNCLOS and the 1993 Convention, Japan put three arguments as to why jurisdiction was wanting. The first was that, Japan having accepted a limit of 1500 tonnes for its experimental fishing programme, the dispute on the tonnage was moot.[187] Secondly, the rest of the dispute was non-justiciable because it related to science rather than law.[188] Thirdly, by entering into the 1993 Convention Japan had fully discharged its

181 Annex VII Tribunal Award (n. 116) 5 [7]. The parties engaged the services of the International Centre for the Settlement of Investment Disputes to provide both the registry for the arbitral tribunal and the venue for its hearings: ibid. [8–9].

182 Ibid. 22–23 [38(a)].

183 Ibid. 25 [38(h)], on art. 281(1), and 27 [39(d)], on art. 283.

184 Ibid. 29–30 [41(a)–(b)].

185 Ibid. 31 [41(d)].

186 Ibid. 32–33 [41(g)].

187 See the argument of Professor Lowe, counsel for Japan, Transcript for 7 May (n. 149) 198–208.

188 See Japanese memorial (n. 149) 83–88 [171–172, 174, 180–181], and the argument of Professor Lowe, counsel for Japan, Transcript for 7 May (n. 149) 181–193.

obligations to Australia and New Zealand in respect of SBT under UNCLOS and its conduct was to be judged solely by reference to the former.[189]

14 The Annex VII Arbitral Tribunal's Award on Jurisdiction

The arbitral tribunal constituted under Annex VII to UNCLOS issued its Award on Japan's objections to its jurisdiction on 4 August 2000. It decided that it lacked jurisdiction to hear the merits of the case and accordingly revoked the provisional measures prescribed by ITLOS. It noted *obiter*, however, that revocation of the ITLOS Order did not mean that the parties might disregard the effects of that Order or their own decisions made in conformity with it. Recalling the parties' statements before it about the progress they had made in narrowing their differences, the Tribunal reminded them that under the 1993 Convention they remained under an obligation to seek to resolve those differences and should refrain from any unilateral act that might aggravate the dispute before it had been fully resolved.[190]

The Annex VII Tribunal rejected Japan's argument that the case was moot: Australia and New Zealand no longer accepted an experimental fishing programme of 1500 tonnes by the time Japan moved to do so, but

> [e]ven if that offer were today accepted, it would not be sufficient to dispose of their dispute, which concerns the quality as well as the quantity of the EFP, and perhaps other elements of difference as well, such as the assertion of a right to fish beyond TAC limits that were last agreed. Japan now proposes experimentally to fish for no more than 1500mt, but it has not undertaken for the future to forego [*sic*] or restrict what it regards as a right to fish on the high seas for Southern Bluefin Tuna in the absence of a decision by the Commission for the Conservation of Southern Bluefin Tuna upon a total allowable catch and its allocation among the Parties.[191]

It did not need to decide the question of the justiciability of scientific matters, which went to admissibility, but observed *obiter* that

189 See Japanese memorial (n. 149) 57–65 [116–131] and the argument of Sir Elihu Lauterpacht, counsel for Japan, Transcript for 7 May (n. 149) 53–58, 63–66, 71–122.

190 Annex VII Tribunal Award (n. 116) 47–48 [67–70].

191 Ibid. 36 [46].

> its analysis of provisions of UNCLOS that bring the dispute within the substantive reach of UNCLOS suggests that the dispute is not one that is confined to matters of scientific judgment only.[192]

On the relationship between UNCLOS and the 1993 Convention, the Tribunal recognised that

> it is a commonplace of international law and State practice for more than one treaty to bear upon a particular dispute.... The current range of international legal obligations benefits from a process of accretion and cumulation; in the practice of States, the conclusion of an implementing convention does not necessarily vacate the obligations imposed by the framework convention upon the parties to the implementing convention.[193]

Expressing scepticism about the consequence of Japan's argument that the obligations of UNCLOS as to SBT no longer bound the parties to the 1993 Convention *inter se* but did still bind them as against all other States Parties to UNCLOS, the Tribunal went on:

> Nor is it clear that the particular provisions of the 1993 Convention exhaust the extent of the relevant obligations of UNCLOS. In some respects, UNCLOS may be viewed as extending beyond the reach of the CCSBT.[194]

It identified a number of UNCLOS obligations not found in the 1993 Convention "operative even where no TAC has been agreed in the CCSBT and where co-operation in the Commission has broken down", which it was prepared to view as in force even where no measures had become binding under the 1993 Convention. Moreover, a dispute concerning the interpretation and implementation of the latter "will not be completely alien to the interpretation or application of UNCLOS for the very reason that the CCSBT was designed to implement broad principles set out in UNCLOS."[195]

On the point on which its decision ultimately turned, the Annex VII Tribunal began its reasoning by finding that the dispute, while centred on the

192 Ibid. 46 [65].

193 Ibid. 40–41 [52].

194 Ibid. 41 [52].

195 Ibid.

1993 Convention, was "a single dispute arising under both Conventions."[196] It then proceeded to determine whether the prerequisites to its jurisdiction under UNCLOS Part XV had been met. Characterising the 1993 Convention as an agreement by the parties to seek settlement of the dispute by a peaceful means of their own choice within the meaning of Article 280, the tribunal next considered whether the requirements of Article 281, paragraph 1 were satisfied. It found the first condition, of no settlement having been reached by the agreed alternative means, met by the "prolonged, intense and serious" negotiations pursued under Article 16 of the 1993 Convention. That provision contained no obligation either to resort in succession to each of the listed means, nor "to negotiate indefinitely while denying a Party the option of concluding for purposes of both Articles 281(1) and 283 that no settlement had been reached."[197] As to the second condition, however, that the agreement "does not exclude any further procedure", the Annex VII Tribunal took the view that Article 16 of the 1993 Convention excluded the dispute settlement procedures of UNCLOS Part XV, so that this condition was not satisfied. Although there was no express exclusion of other procedures, the tribunal found this was not necessary for the purposes of Article 281, paragraph 1. Rather, since under Article 16 the consent of all parties to the dispute was needed for it to be heard by an arbitral panel or the ICJ, in the absence of such consent they were obliged to continue to seek resolution of the dispute only by the peaceful means there listed, to the exclusion of all other mechanisms. Emphasising the consensual nature of the dispute settlement process under the 1993 Convention, despite there being no express exclusion of any procedure, the Annex VII Tribunal read into Article 16 an intention on the part of its parties for this to be the sole avenue for settling disputes *inter se* concerning SBT, so removing such disputes from the compulsory processes of UNCLOS.[198] Reinforced by the observation that because substantial categories of disputes are or may be excluded by Section 3 of Part XV, "UNCLOS falls significantly short of establishing a truly comprehensive regime of compulsory jurisdiction entailing binding decisions",[199] the tribunal concluded that Article 16 accordingly excluded "any further procedure within the contemplation of Article 281(1) of the LOS Convention."[200] As a result, it held that it had no jurisdiction to proceed to the merits of the

196 Ibid. 41–42 [52], [54].

197 Ibid. 42–43 [55].

198 Ibid. 43–44 [57]. As the Annex VII Tribunal saw it, the fact that the 1993 Convention included an annex detailing its own arbitral process fortified its reasoning.

199 Ibid. 45 [62].

200 Ibid. 44 [59].

case[201] and did not need to consider the competing arguments as to admissibility of the claims, hence its Award expressed no view on them. Significantly, however, the Award not only did not criticise the prior ITLOS decision granting provisional measures, but went out of its way to say that their effect had been useful in bringing the parties' positions closer together.[202]

Possibly to the applicants' chagrin,[203] there is a hint in the Award that the result might have been different had they invoked Article 300 in its own right:

> The Tribunal does not exclude the possibility that there might be instances in which the conduct of a State Party to UNCLOS and to a fisheries treaty implementing it would be so egregious, and risk consequences of such gravity, that a Tribunal might find that the obligations of UNCLOS provide a basis for jurisdiction, having particular regard to the provisions of Article 300 of UNCLOS. While Australia and New Zealand in the proceedings before ITLOS invoked Article 300, in the proceedings before this Tribunal they made clear that they do not hold Japan to any independent breach of an obligation to act in good faith.[204]

Sir Kenneth Keith dissented from this analysis. His view was that Article 16 of the 1993 Convention was not an agreement to resolve disputes by another method for the purposes of UNCLOS Article 281, paragraph 1, since it did not prescribe any particular method. In any event, he found that Article 16 only applied to disputes concerning the 1993 Convention and did not necessarily apply to disputes concerning UNCLOS.[205] Despite the central place within UNCLOS of the compulsory dispute settlement scheme, to which the exceptions were well defined, Sir Kenneth was prepared to concede that Article 16 could in theory have excluded UNCLOS Part XV as among the parties to both treaties, but in the absence of "strong and particular wording" to that effect, this was not the case.[206]

201 Ibid. 46 [65].

202 Ibid. 46–47 [65], [67].

203 See (n. 5).

204 Annex VII Tribunal Award (n. 116) 46 [64].

205 *Southern Bluefin Tuna* (*Australia v Japan; New Zealand v Japan*) (Award on Jurisdiction and Admissibility, Sep. Op. Keith) (2000) XXIII *RIAA* 49, 51 [8]. Sir Kenneth added ibid. [12]: "That the disputes may or may not also concern the interpretation or implementation of the [1993 Convention] is beside the point."

206 Ibid. 53–56 [18], [21]–[29].

The Award has been widely criticised,[207] most forcefully by Colson and Hoyle, who call for the Award not to be regarded as a precedent but instead to be limited to the specific facts of the case,[208] describing its interpretation of Article 281, paragraph 1 as "questionable" and "based on a view of compulsory dispute settlement under the LOS Convention that is not widely shared."[209] They attack the tribunal's interpretation of Article 281, paragraph 1 as permitting compulsory dispute settlement under UNCLOS to be defeated by consensual arrangements even where there is no clear manifestation that the parties intended to contract out of it. This they regard as "substantially not in keeping with the intent of the negotiators of [UNCLOS]", whose aim had been "to establish a comprehensive dispute settlement framework relating to activities in the world's oceans", with exceptions from the compulsory system for certain issues having to be conceded via Section 3 of Part XV as the necessary price of "achiev[ing] international consensus on the entire Convention package." Just as importantly, certain other classes of issues "were not excluded by Section 3, including disputes concerning living resources beyond coastal state jurisdiction (and southern bluefin tuna is certainly in this category)[.]"[210]

Additionally, Colson and Hoyle draw attention to the provisional measures phase of the *MOX Plant* case heard a year later by ITLOS, which, they conclude, appears to have indirectly rejected the reasoning of the *Southern Bluefin Tuna* award.[211] Although the MOX Plant dispute as a whole between Ireland and the United Kingdom involved several treaties, ITLOS found that there was *prima facie* jurisdiction under Part XV of UNCLOS; in particular, the dispute before itself and the Annex VII Tribunal concerned UNCLOS alone.[212] Similarly, ITLOS expressed a narrow view of Article 282, restricting it to agreements that provide for dispute settlement with regard to the "interpretation or application

207 See e.g. Alan Boyle, 'Problems of Compulsory Jurisdiction and the Settlement of Disputes Relating to Straddling Fish Stocks' in Olav Schram Stokke (ed), *Governing High Seas Fisheries: The Interplay of Global and Regional Regimes* (Oxford University Press 2001) 91, 104–105; Oxman (n. 3).

208 Colson and Hoyle (n. 3) 70.

209 Ibid. 76. They highlight at 70 the perversity of the outcome for Australia and New Zealand, if compulsory dispute settlement were the sole consideration, that they "would have been better off not having a CCSBT to address the specifics of southern bluefin tuna conservation and management." Accord the submissions to the Annex VII Tribunal of counsel for Australia, Professor Crawford, Transcript for 8 May (n. 5) 139–140.

210 Colson and Hoyle (n. 3) 67.

211 Ibid. 72–74.

212 MOX Plant Case Order (n. 171) 106 [52].

of this Convention."[213] The test of whether the dispute settlement mechanism under another agreement involves an interpretation or application of UNCLOS is whether it is capable on its face of resolving the UNCLOS dispute; the subject matter of the agreement is irrelevant. Thus ITLOS declined to follow the SBT case, in which the Annex VII Tribunal had found that the same dispute arose under both UNCLOS and the 1993 Convention, deciding instead that the matter before it was a dispute under UNCLOS that other agreements did not regulate unless either they mentioned UNCLOS by name or it fell into a specifically mentioned class of treaties.[214]

In the *South China Sea arbitration* between the Philippines and China which came before a differently composed Annex VII Tribunal, a "position paper" released by China had adopted the *Southern Bluefin Tuna* Tribunal's reasoning on Article 281 in relation to the 2002 ASEAN-China Declaration on the Conduct of Parties in the South China Sea,[215] which it portrayed as an agreement to resolve disputes relating to the South China Sea solely through negotiation.[216] The tribunal treated the Chinese position paper as an objection to its jurisdiction and held a separate preliminary hearing on this matter.[217] The Philippines

213 Ibid. [48].

214 Ibid. [53]. Several judges delivered Separate Opinions in which they suggested that ITLOS would be right to depart from the Annex VII Tribunal's reasoning regarding art. 281(1) in the SBT case; Judge Wolfrum, for example, observed (*MOX Plant Case* (*Ireland v. United Kingdom*) (Provisional Measures, Order of 3 December 2001, Sep. Op. Wolfrum) [2001] ITLOS Rep 131, 131–132) that the United Kingdom's interpretation of art. 282 defeated the objective of Part XV of UNCLOS. Since more than one treaty might bear upon a single dispute, which could have diverse outcomes depending upon what treaty were invoked, "an intention to entrust the settlement of disputes concerning the interpretation and application of the Convention to other institutions must be expressed explicitly in respective agreements."

215 ASEAN-China Declaration on the Conduct of the Parties in the South China Sea (4 November 2002) <http://www.asean.org/?static_post=declaration-on-the-conduct-of-parties-in-the-south-china-sea-3&category_id=32> accessed 2 April 2016.

216 Ministry of Foreign Affairs of the People's Republic of China, 'Position Paper of the Government of the People's Republic of China on the Matter of Jurisdiction in the South China Sea Arbitration Initiated by the Republic of the Philippines' (7 December 2014) [38–40], [81–82] <http://www.fmprc.gov.cn/mfa_eng/zxxx_662805/t1217147.shtml> accessed 2 April 2016.

217 *In the Matter of an Arbitration before an Arbitral Tribunal Constituted under Annex VII to the 1982 United Nations Convention on the Law of the Sea* (*Republic of the Philippines v. People's Republic of China*) (Award on Jurisdiction and Admissibility, Order of 29 October 2015) (hereinafter South China Sea Arbitration Award on Jurisdiction) PCA Case No. 2013–19, 22–24 [64], [68] <http://www.pcacases.com/web/sendAttach/1506> accessed 2 April 2016.

invited the tribunal to follow not the reasoning of the majority in the earlier case, but rather that of Sir Kenneth Keith in the minority, which it argued was the better test (i.e. did the Declaration expressly exclude any further proceedings under UNCLOS?).[218] Although the tribunal expressed agreement with this argument,[219] the force of this was diminished because it took the form of an *obiter dictum*, as it had already found that this instrument was not of treaty status and could not therefore engage Article 281.[220]

15 Resolution of the Dispute

a *The SBT Scientific Research Program*

Following the Award, Japan advised Australia and New Zealand that it wished to see a return to consensus and cooperation in the CCSBT. Proposing high-level negotiations for that purpose, Japan indicated that it did not intend to conduct any further experimental fishing unilaterally. In the subsequent negotiations it was agreed that the way to resolve the disagreement about the appropriate nature and extent of any experimental fishing was to engage independent external scientists to devise a scientific programme to reduce the uncertainties in relation to the SBT stock.[221] At a special meeting of the CCSBT in November 2000, the parties adopted terms of reference for the external scientists and agreed that whatever scientific programme the external scientists recommended would become the CCSBT's decision, unless there was agreement to alter it.[222]

218 Oral submissions of Counsel for the Philippines, Mr Martin, *In the matter of an arbitration under Annex VII to the United Nations Convention on the Law of the Sea* (Hearing on Jurisdiction and Admissibility, Day 2) PCA Case No. 2013–19, 13–16 <http://www.pcacases.com/web/sendAttach/1400> accessed 2 April 2016. The written submissions of the Philippines, though referred to in the Award, had not yet been made public at the time of writing.

219 South China Sea Arbitration Award on Jurisdiction (n. 217) 86–88 [221–229].

220 Ibid. 82–85 [212–218].

221 Mansfield (2001) (n. 3) 365.

222 See 'Development of a SBT Scientific Research Program including a Scientific Fishing Component by the CCSBT External Scientists' (Attachment L to CCSBT, *Report of the Special Meeting* (16–18 November 2000) (hereinafter CCSBTSM3 Report, 3 <http://www.ccsbt.org/userfiles/file/docs_english/meetings/meeting_reports/ccsbt_06/report_of_special_meeting.pdf> accessed 10 August 2015)).

At its next meeting in April 2001, the CCSBT adopted the scientific research programme developed by the independent scientists.[223] Providing for a wide range of research activities to improve understanding of the SBT stock, the programme centred on work to determine the actual catch of SBT, to develop a more effective stock assessment model and a fish tagging programme to provide a better indication of stock levels. It also contemplated the placing of observers on SBT fishing vessels to monitor their catch.[224]

Although it would be an exaggeration to say that this resolved for all time disagreements about the measures necessary to manage the SBT stock, as a degree of difficulty and controversy is inevitable about TAC and national allocations, the agreed scientific research programme in time produced a much improved basis for decision-making. This culminated in the adoption of a management strategy at the CCSBT's 2005 meeting,[225] although for other reasons the strategy could not be implemented and was ultimately superseded in 2011 by a replacement strategy.[226]

b *Japanese Commercial Catch Affected by the ITLOS Order*

The Annex VII Tribunal's lifting of the ITLOS provisional measures led to renewed controversy over how to treat the 711 tonnes of commercial catch Japan had forgone in part-compliance with the two-year catch limit under that Order: was it entitled to reclaim it, or was the situation covered by the tribunal's dictum on not disregarding the effects of decisions made in conformity with the Order?[227] In the end, as part of the settlement of the dispute Australia

---

223 CCSBT7 Report (n. 118) [12].

224 See *Report of the SC to the CCSBT on the Scientific Research Program* (Attachment D to CCSBT, *Report of the Fifth Meeting of the Scientific Committee* (19–24 March 2001) <http://www.ccsbt.org/userfiles/file/docs_english/meetings/meeting_reports/ccsbt_07/report_of_sc5.pdf> accessed 10 August 2015).

225 CCSBT, *Report of the Extended Commission of the Twelfth Annual Meeting of the Commission* (11–14 October 2005) (Appendix 3 to CCSBT, *Report of the Twelfth Annual Meeting of the Commission* (15 October 2005) [69] <http://www.ccsbt.org/userfiles/file/docs_english/meetings/meeting_reports/ccsbt_12/report_of_ccsbt12.pdf> accessed 10 August 2015).

226 See Resolution on the Adoption of a Management Procedure (Attachment 12 to CCSBT, *Report of the Extended Commission of the Eighteenth Annual Meeting of the Commission* (10–13 October 2011) (Appendix 3 to CCSBT, *Report of the Eighteenth Annual Meeting of the Commission* (10–13 October 2011) <http://www.ccsbt.org/userfiles/file/docs_english/meetings/meeting_reports/ccsbt_18/report_of_CCSBT18.pdf> accessed 10 August 2015)).

227 Annex VII Tribunal Award (n. 116) 47 [67].

and New Zealand agreed that in 2001 Japan would be able to add half of this amount, or 356 tonnes, to its commercial catch.[228]

c *The New-Old Quota of 2003*

The Scientific Committee at its next meeting in August 2001 unanimously concluded that the risk of further recruitment declines, while not possible to determine quantitatively, was not particularly high, and that "[a]n immediate reduction in total removals is thus not recommended as a necessary action to prevent stock collapse ... a policy of maintaining current removals would most likely enable the CCSBT to react in a timely fashion to future stock trends." Nonetheless there was "a risk of further stock declines if current removals are maintained, and depending upon members['] aversion to this risk, differing levels of catch reductions would be appropriate forms of insurance for the sustainability of the current fishing industries."[229]

Even so, this was not sufficient to prevent a continuation of the pattern in quota negotiations when the report was considered by the Eighth Meeting of the CCSBT. Japan again demanded an increase, this time of 500 tonnes to be shared *pro rata* among Members, which was resisted by the other parties, resulting once more in no TAC and national allocations being set.[230] Even though agreement was reached in principle on a "provisional catch limit based on current catch levels",[231] ambiguity remained as to what the "current" level actually meant, owing to uncertainty surrounding the Indonesian catch, which prevented any agreement on national allocations[232] and thus in turn on a TAC as their sum. The same outcome occurred at the Ninth Meeting in 2002, where Japan stated that for setting the TAC "account needed to be taken of both the Scientific Committee's recommendation and socio-economic issues". It believed that, even if the TAC were set at the level of the 2001 catch, there

228 For criticism see Andrew Serdy, 'The Paradoxical Success of UNCLOS Part XV: A Half-Hearted Reply to Rosemary Rayfuse' (2005) 36 *Victoria University of Wellington Law Review* 713, 716n and source there cited.

229 CCSBT, *Report of the Sixth Meeting of the CCSBT Scientific Committee* (28–31 August 2001) (hereinafter CCSBT-SC6 Report) <https://www.ccsbt.org/sites/ccsbt.org/files/userfiles/file/docs_english/meetings/meeting_reports/ccsbt_08/report_of_sc6.pdf>, accessed 25 August 2016, 12 [29–30].

230 CCSBT, *Report of the Eighth Annual Meeting* (15–19 October 2001) (hereinafter CCSBT8 Report) 10–11 [70–81] and Attachment N-2 ("Statement by Japan on Agenda Item 9.2") <https://www.ccsbt.org/sites/ccsbt.org/files/userfiles/file/docs_english/meetings/meeting_reports/ccsbt_08/report_of_ccsbt8.pdf> accessed 25 August 2016.

231 Ibid. 11 [75].

232 Ibid. [81].

would be a surplus within it that could be reallocated to existing Members' national allocations.[233]

Agreement drew closer in April 2003 with the conclusion of participants at the Indonesian Catch Monitoring Review Workshop that, with improvements in dockside monitoring having resolved many of the uncertainties, "current Japanese import data should not be used to estimate total Indonesian SBT catch from monitored landings."[234] Finally, at the Tenth Meeting a TAC of 14,030 tonnes was adopted, divided into national allocations as follows:

Australia	5265
Japan	6065
Korea (Republic of)	1140
New Zealand	420
Taiwan (Fishing Entity of)	1140
TOTAL	14030[235]

It will be noted that the national allocation of each original party to the 1993 Convention was exactly the same as under the last TAC set in 1997, the level for which Australia and New Zealand had consistently argued in the intervening years but which Japan had rejected. It is ironic that, despite the belated acknowledgment by Japan of the impossibility of meeting the 1980-by-2020 target for parental biomass, their position ultimately prevailed for an entirely unrelated reason: the glut of SBT on the Japanese market.[236]

233 CCSBT, *Report of the Extended Commission of the Ninth Annual Meeting of the Commission* (15–18 October 2002) (Appendix 3 to CCSBT, *Report of the Ninth Annual Meeting of the Commission* (15–18 October 2002) 24 [77], [80] <http://www.ccsbt.org/userfiles/file/docs_english/meetings/meeting_reports/ccsbt_09/report_of_ccsbt9.pdf> accessed 20 August 2015).

234 CCSBT, *Report of the Indonesian Catch Monitoring Review Workshop* (10–11 April 2003) 5 [19] <http://www.ccsbt.org/userfiles/file/docs_english/meetings/meeting_reports/ccsbt_10/report_of_icmws.pdf> accessed 10 August 2015.

235 All figures are in tonnes. Additionally a collective allocation of 900 tonnes was made to cooperating non-members, of which 800 tonnes would be offered to Indonesia: CCSBT-EC Report 2003 (n. 150) 10 [51].

236 Text at (n. 150).

16 Issues Left Unresolved Because the Case did not go to the Merits

a *Possible Damaging Admissions by Japan in the Indian Ocean Tuna Commission*

Although Japan presented the results of its first year of experimental fishing to the First Session of the Scientific Committee of the Indian Ocean Tuna Commission (IOTC),[237] whose constitutive convention also gives it competence to regulate SBT,[238] it did not present the results for the second year.[239] While one explanation for this might be a wish on Japan's part to respect the terms of one of the ITLOS orders, that each party to the dispute "ensure that no action is taken which might extend or aggravate the disputes submitted to the arbitral tribunal",[240] another possible reason is that the more rigorous design of the second year's experiment produced results that were not as helpful to its case for increase in catch as those of the first year.

Oddly, given the fact that the CCSBT Executive Secretary's written report to the IOTC would have been cleared by it, Japan did not object to the gloss put by that document's description on the continuing impasse in relation to the TAC, which was adopted unchanged in the report of the meeting: "No agreement has been reached in relation to the Total Allowable Catch (TAC), which remains at the 1997 levels."[241] This directly supports the contention of Australia and New Zealand in the dispute that, in the absence of a CCSBT decision on a

---

237 IOTC doc IOTC/S/03/98/R[E], *Report of the Third Session of the Indian Ocean Tuna Commission* (9–12 December 1998) 19 <http://www.iotc.org/sites/default/files/documents/proceedings/1998/s/IOTC-1998-S03-R%5BEN%5D.pdf> accessed 18 August 2015 (Appendix G, *Report of the First Session of the Scientific Committee* 33 (Appendix E, 'Preliminary result of Experimental Fishing Program (EFP) of Southern Bluefin Tuna (SBT) conducted by Japan')).

238 Agreement Establishing the Indian Ocean Tuna Commission (concluded 25 November 1993; entry into force 27 March 1996) 1927 UNTS 329, art. III and Annex B (item 5).

239 The results were eventually presented to the CCSBT: the reference in 'Opening statement by Japan' (Attachment D to CCSBTSM3 Report (n. 222)) suggests that this was done in 2000.

240 ITLOS Order (n. 159) 297 [90.1(a)].

241 IOTC doc IOTC/S/04/99/R[E], *Report of the Fourth Session of the Indian Ocean Tuna Commission* (13–16 December 1999) 5 [38] <http://www.iotc.org/sites/default/files/documents/proceedings/1999/s/IOTC-1999-S04-R%5BEN%5D.pdf> accessed 5 August 2015.

TAC, the parties were obliged to stay within their most recent national allocations and is incompatible with Japan's contrary position.[242]

b *Best Scientific Evidence (Design of the Experimental Fishing Programme and Analysis of Its Results)*

Given the failure of the depleted SBT stock to recover,[243] Australia and New Zealand argued that the flaws in the design of the experiment itself, and of the way in which Japan proposed to use its results in future stock assessments, deprived it of scientific value and were outweighed by the significant risk posed to the stock by the greater catch. As to the design, it will be recalled that the Japanese experimental fishing sought to reduce the uncertainty in the interpretation of CPUE by collecting catch and effort data from area/month combinations in which no recent commercial fishing had taken place, in order to estimate the ratio of relative density of SBT in 5°-square areas not fished commercially to the density in areas that were continuing to be fished. In itself, this is unobjectionable: according to Polacheck,[244] the relative density in an area and season in one year does not provide a reliable basis for estimating the relative density in that area and season in other years. For CPUE to provide a statistically valid index of abundance, it would be necessary to account for the relative density of fish in areas and periods with no fishing.[245] In Polacheck's

---

242 In their Statement of Claim (n. 5) [69(1)(b)], one of the declarations sought by Australia and New Zealand from the Annex VII Tribunal was that Japan had breached its obligations to them by "carrying out unilateral experimental fishing in 1998 and 1999 which has or will result in SBT being taken by Japan over and above previously agreed Commission national allocations".

243 Despite the 1989 reduction in the TAC (text following n. 106), the size of the spawning stock at the time of the dispute was between 25% and 53% of the 1980 level. Relative to stock levels in the fishery's early years, Australian scientists estimated it at historically low levels in the order of 7–15% of its 1960 level, and Japanese scientists at 12% of the 1951 level, with recruitment (defined (n. 77)) having declined markedly from the late 1960s to the mid-1990s to around a third of the 1960 level: see Statement of Claim (n. 5) [6]. Polacheck (n. 3) 292 states that the decades of intensive fishing of the SBT stock had resulted in a population that was significantly overfished and below commonly accepted thresholds for biologically safe parental biomass, beneath which the risk of poor recruitment increases. The greatest concern was that natural environmental variability could combine with the vulnerable state of the resource to cause abrupt recruitment decline and a subsequent (and consequent) further decline in the parental stock. This had been the mechanism associated with a large number of fisheries collapses.

244 Polacheck (n. 3) 287.

245 'Status of the stock and fishery indicators' (Appendix 1 to 12th Trilateral Scientific Meeting Report (n. 112)) 1 [1] cautions that, although CPUE is of great value in indicating general

view, there were numerous problems with Japan's proposed methods for analysing and interpreting results of the experimental fishing. These included the fact that the ratio concerned was unlikely to be constant spatially, temporally or across all age classes, confounding any extrapolation to other areas and past years. Moreover, estimates of the ratio would depend on their spatial and temporal scales, the criteria used to define fished and unfished areas and the model used to calculate SBT densities in the area not fished in the experiment. From the 1998 results, very different values for the ratio could be obtained, depending on how these factors were handled.[246]

Further, Japan's experimental design had no mechanism to ensure that sampling was distributed representatively throughout the area and time period of the experiment. The vessel deployment scheme did not constrain where half the vessels fished, and the other half had only minimal constraints. As a consequence, the actual 1998 experimental effort was highly concentrated in or very close to the commercially fished area, only 3% of the total effort in the experiment occurring 120 nautical miles or more—that is, twice the length of an average longline—from the commercially fished area (see Map 1).

More bizarrely yet, in the process of refinement from the original proposal for experimental fishing, the justification given by Japan for the experiment that finally took place was simplified to the point of absurdity. The simplified reason was to disprove the variable squares hypothesis. This would have been open to legal attack in two obvious ways and a more subtle one. First, Australia and New Zealand had never asserted the hypothesis as a reflection of reality, even though their models that accorded it some weight appeared to have predictive power (a fact that would not have changed even if the hypothesis were affirmatively shown to be false). Secondly, even if Australia and New Zealand had maintained the hypothesis, an experimental catch of 6,000 tonnes over three years would not have been necessary to refute it; the catching of a single SBT in an unfished square for the relevant month would have sufficed. Thirdly, if, as Japan proposed, the variable squares hypothesis had to be removed from the models as a result of being disproved, the effect of the remaining hypotheses would by definition have been to shift the prognoses for the stock towards the optimistic end of the spectrum—and this irrespective of the actual state of

trends, interpretation of relatively small and short term changes is very difficult because many factors can change CPUE other than abundance, especially for small fish; the data available have limited capacity to correct for their effects.

246 Polacheck (n. 3) 289–290.

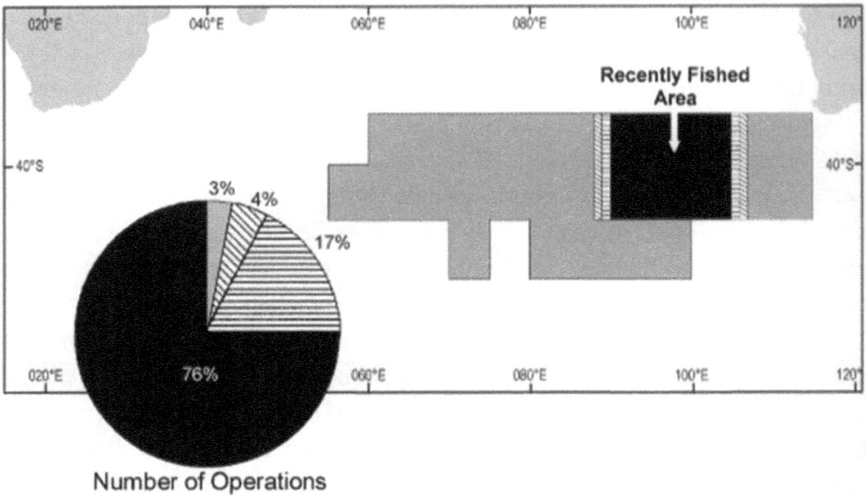

MAP 1 *The area for which the Japanese 1998 experimental fishing intended to estimate the relative density of* SBT *in fished and unfished 5° squares and the relative amount of effort within different portions of this area.*[247]

the stock disclosed by the experiment.[248] This is no more sensible than trying to settle a dispute about the ratio in which black and white paints have been mixed to produce a particular shade of grey by proving that the mixture is not pure white, in order to insist that it must be "blacker than we thought", a manifest *non sequitur*.

247 Taken from ibid 290, reproducing a graphic in the public domain presented to ITLOS by Professor Crawford, Counsel for Australia, referred to as "Tab 13" in Transcript for afternoon of 18 August (n. 73), 21. Shaded black are the 5° squares fished regularly in the 1990s. The adjacent vertical strips with horizontal and diagonal lines represent the area within 1° and 2° respectively of those fished squares, while the grey region represents the rest of the area.

248 Note also Polacheck's argument, (n. 3) 289 that, although the relative weights given to the different CPUE models contributed to the differences in the parties' estimates of the probability of recovery, they were not the principal source of differences in the projection estimates. The weights assigned to the different CPUE models by Australia and Japan were in fact quite similar, and the estimated probabilities from the 1998 assessment were largely insensitive to them. Moreover, even if the constant squares hypothesis, which was the most optimistic interpretation of CPUE, could be shown to be correct, the Australian scientists' estimate of the recovery probability from the 1998 assessment was still only 36% and it was 40% for their New Zealand counterparts.

For all the understandable reluctance of legal tribunals to evaluate competing scientific judgements, if the applicants had succeeded in proving the facts recited in the previous paragraph, it is not obvious how the Annex VII Tribunal could have stopped short of drawing the necessary conclusion that the experiment itself, and any measures said to flow from its results, would have failed to meet the standard of being based on the "best scientific evidence available" laid down in UNCLOS Article 119, paragraph 1(a). Japan would, it seems, have found it necessary to argue that "best… available" means "best that the commercial fleet can be persuaded to gather". Commercial catch needs would thus be acting as a constraint on scientific ones rather than *vice versa*. Similar considerations came to the fore in the *Whaling* case: despite having gone to some lengths to work out how many whales needed to be caught in order for its sampling programme to deliver meaningful results, Japan made no attempt either to remedy the often considerable shortfalls that occurred year after year or to redesign the programme, a point to which the ICJ attached much significance.[249]

As a matter of law, it is hard to disagree with Polacheck's conclusion that catch taken in the name of science cannot be an end in itself. Had the SBT case gone to the merits, the main issue would have been the standard and burden of proof required for experimental fishing. Polacheck sees the precautionary approach in Annex II to the UN Fish Stocks Agreement[250] (which does not apply to whales) placing the onus on the proponent of a course of action involving increased risk. More specifically, could substantial increases in catches be justified on scientific grounds in the absence of agreement on the validity of the experimental design to provide meaningful results and of any specific management framework in which those results would be used, and without demonstrating that the experiment, even if conclusive, would substantially change the existing scientific advice?[251]

An interpretation of the precautionary approach to require favouring a conservative measure over a risky one may be too simplistic: this is possibly sufficient for depleted stocks, but it may not be appropriate to give it normative

249 *Whaling in the Antarctic* (n. 1) 293 [226].

250 Agreement for the Implementation of the Provisions of the United Nations Convention on the Law of the Sea of 10 December 1982 relating to the Conservation and Management of Straddling Fish Stocks and Highly Migratory Fish Stocks (opened for signature 4 December 1995, entered into force 11 December 2001 according to the UN depositary website, though see Andrew Serdy, "How Long has the UN Fish Stocks Agreement Been in Force?", (2003) 34 *Ocean Development & International Law* 29) 2167 UNTS 3.

251 Polacheck (n. 3) 293.

character for all stocks. Where scientific modelling is at issue, a compound condition suggests itself: if one model robust to error in assumptions favours conservation, and a less robust one does not (robust in the sense that slightly different assumptions generate slightly, not significantly different outcomes), then the law too should support the robust model.[252]

Although one can accept Japan's view that "[e]xcessive conservatism of not accepting any additional risk under any circumstance will be adverse for the achievement of maximum sustainable use of resources", which Japan regarded as "clearly supported" in the 1993 Convention,[253] if this was intended to be a reference to the Australian and New Zealand position, then it is no more than a caricature of it. The latter States' position was not one of unwillingness to accept any additional risk at all, but of accepting it only if it could be shown to produce worthwhile benefits.

c *The Allegation that Japan's Experimental Fishing was Commercially Motivated*

The issue of commercial motivations for ostensibly experimental or scientific activity was common to both disputes, but the parallels were not exact. In the *Whaling* case, two of the three specific provisions of the Schedule to the ICRW invoked by Australia did refer to "commercial whaling": the zero catch limit in paragraph 7(b) and paragraph 10(e) establishing the Southern Ocean Sanctuary, but not the prohibition on use of factory ships in paragraph 10(d). Yet it was not necessary for Australia to prove that Japan's whaling was driven by commercial rather than scientific motives. This was because it was common ground between the parties that the applicable law, Article VIII of the ICRW, did not create a third category of whaling, rather all whaling was simply regarded as being for commercial purposes unless it could be shown to be "for purposes of scientific research" within the meaning of its paragraph 1. The ICJ too took the view that all three paragraphs were intended to cover all killing, taking and treating of whales that was not for such purposes (other

252 Note that the Japanese models were indeed less robust: Japan's estimated probability in 1997 of recovery of the parental stock to its 1980 level by 2020 was 79%, as against 36% for Australia and 29% for New Zealand, but additional Japanese analyses with a "slightly different" virtual population analysis structure (see Appendix infra) gave 20%: CCSBT3(1) Report (n. 118) 16. In Australia's view this indicated that the state of those analyses given high weights by Japan (only 24 of the 216 agreed on by the Scientific Committee), and thus the projection results based on only 12 of these 24, did not provide a robust measure of the status of the stock. Japan excluded the other 192 because it considered these to be outside the range of plausible hypotheses, although it was alone in this view: ibid. 19.

253 Ibid. 18.

than aboriginal subsistence whaling under paragraph 13 of the Schedule, which by its very nature would not entail any of these activities). The Court rejected the implication that there could be other kinds of whaling not authorised by either Article VIII, paragraph 1 of the ICRW or paragraph 13 of the Schedule but nonetheless falling outside the scope of the prohibitions at issue; any such interpretation would undermine the object and purpose of the ICRW by leaving these undefined categories of whaling activity beyond its scope.[254] Australia pointed to the quantity of whale meat sold as calling into question whether the whaling was truly for scientific purposes,[255] but the ICJ concluded that this alone should not have that effect, noting that the ICRW in Article VIII, paragraph 2 expressly permitted the sale of meat.[256] It is suggested that Japan's motives are better analysed as financial rather than commercial: even if the underlying purpose of JARPA II was to show that it would be safe to resume commercial whaling, the programme itself ran at a loss which the proceeds of meat sales served to reduce, and the wish to avoid compounding the loss would explain, if not excuse, the absence of redoubled efforts to catch the full quota of 850 minke whales annually when obstacles to this were encountered.

In the *Southern Bluefin Tuna* arbitration, by contrast, there was no prohibition on commercial fishing, a practice in which all three parties continued to engage. Instead, the allegation of commercial motives[257] served to attack the

---

254 *Whaling in the Antarctic* (n. 1) 294 [229]–[230].

255 Ibid. 259 [91].

256 Ibid. [92], [94].

257 The Australian Minister responsible for fisheries had issued a press release attacking the experimental fishing programme as a "pretext to increase Japan's catch": Japanese memorial (n. 149) 29 [61]. In their pleadings Australia and New Zealand continued to maintain that the programme was wholly or partly disguised commercial fishing: see e.g. the Statement of Claim (n. 5) [48(d)], noting that, "although the ostensible aim of the program was to determine the density of SBT in areas formerly but no longer fished, the proposed vessel deployment scheme would result in a substantial majority of the effort and catch being taken from time and area strata still being fished, on an essentially commercial basis"; *Southern Bluefin Tuna Cases (New Zealand v. Japan; Australia v Japan)* (Request for provisional measures, Oral submissions to ITLOS of the Australian Attorney-General, Mr Williams QC) ITLOS doc ITLOS/PV.99/20/Rev.2 14–15 <https://www.itlos.org/fileadmin/itlos/documents/cases/case_no_3_4/VRE180899am.corr-rev2.pdf> accessed 12 August 2015 (hereinafter Transcript for morning of 18 August), as well as those of the Agent for Australia, Mr Campbell, ibid. 20–22, Counsel for New Zealand, Mr Mansfield, ibid. 30, Counsel for Australia, Professor Crawford, Transcript for afternoon of 18 August (n. 73) 18 and Counsel for Australia, Mr Burmester, ibid. 32, 34–35; *Reply on Jurisdiction [of] Australia and New Zealand* (n. 154) 98 [A34]: Japan's final proposal was "accompanied by papers setting out the objectives of the proposal and views by Japanese industry

legitimacy of Japan's experimental fishing programme, so here too it may be thought unlikely that the Annex VII Tribunal would have found it necessary to make any findings on the matter. While there was a strong reason to suspect such motives given the way Japan had presented its initial proposals for experimental fishing in 1995,[258] there are a number of indications that, by the time the fishing was actually carried out, any hope of profit from it had vanished. Japan's pleadings refer on numerous occasions to the losses the experiment entailed for its fleet,[259] and while Australia and New Zealand treated this sceptically, as it would be easy to conceal profits through creative accounting, it is suggested that there was probably a good deal of truth to Japan's protestations. This is because the changes to the programme made as a consequence of the two years of negotiations, while not enough to satisfy the demands of Australia and New Zealand regarding the scientific merit and rigour of the experiment, had put sufficient constraints on the Japanese fleet in terms of when and where they fished to make it improbable that the venture would be a profitable one for them. If so, this also provides a rational explanation for one otherwise inexplicable aspect of the conduct of the Japanese fisheries authorities: their repeated assertions that Australia and New Zealand actually believed the variable squares hypothesis,[260] despite the applicants' clear denial of this.[261] The assertion appears to have been regarded by the Japanese

---

which made it clear that the EFP was designed to deliver a profit to Japanese fishers rather than scientifically meaningful results"; Transcript for 8 May (n. 5) 22–23 (per the Agent for Australia, Mr Campbell).

258 See Transcript for morning of 18 August (n. 257) 20 and 24–25 (per the Agent for Australia, Mr Campbell).

259 Japan's experimental fishing programme was "not commercially viable, [having to be undertaken] at significant expense to [Japan] and to its fishing industry" (Japanese written statement to ITLOS (n. 131), [40]); Transcript for morning of 19 August (n. 103) 9 (per the Agent for Japan, Mr Togo); Transcript for 7 May (n. 149) 41 (per Counsel for Japan, Professor Ando).

260 See the oral presentation by Mr Greig, counsel for Japan, before ITLOS, Transcript for morning of 19 August (n. 103) 19:

"Applicants are also assuming that there are no fish even in an area commercially fished, in any month in which there is no commercial fishing in that area. In effect, they are assuming that all the southern bluefin tuna in a given 5 degree by 5 degree area swim away as soon as the commercial fishing ends. In other words, area 8, which is the area of commercial fishing shown, is thought to be devoid of fish in months when the fishing is out of season. That is the reason you have to do some experimental fishing during those months to test that extreme assumption."

261 Transcript for afternoon of 18 August (n. 73) 19 (per Counsel for Australia, Professor Crawford).

authorities as necessary to persuade their own industry of the need for the experiment, in which they had become reluctant to take part, fearing that they would incur financial loss through the lower catch rates they expected in the unfished squares where they would have to fish.[262] (Though credible in itself as a factor tending to refute the allegations of commercial motives, it may be observed that this argument necessarily casts severe doubt on the constant squares hypothesis that Japan favoured.)

It is noteworthy that Morgan, a member of Japan's legal team, later opined that, if the stock were truly in danger, scientific catch should have taken precedence over commercial catch.[263] This is defensible, but was not his client's position—indeed, it was closer to the Applicants': that if Japan considered the experiment so vitally important, it was at liberty to conduct it unilaterally, provided it did so within its last agreed national allocation, i.e. at the expense of its own commercial catch. (In fact Japan had done just that some years earlier, adding 40 tonnes of its own allocation to a trilateral research quota because it believed "the programme was necessary in order to understand the real state of the stock".[264]) On the need for additional catches as opposed to accommodating the experimental fishing within Japan's previous national allocation, Polacheck argues that this need was never demonstrated in terms of the information supposedly to be gained; rather it was dictated by "political/economic realities". If the same information could be obtained without increasing catches, the increased short-term risks of additional catches would not be justifiable simply for their potential to decrease longer-term risks. Agreeing with Morgan, Polacheck states that if, on the other hand, some form of experimental fishing or scientific catch were required to provide the information for management or for reducing risk, then such catch should take priority, commercial catches being added only if the recovery strategy leaves scope for them.[265]

---

262 See e.g. Japan Fisheries Association, 'Southern Bluefin Tuna Experimental Fishing Program for 1999' <http://www.suisankai.or.jp/iken_e/iken99_e/ik002_e.html> accessed 22 August 2015, containing the passage:

"The first year of the Experimental Fishing Program confirmed that Australia and New Zealand's hypotheses were unrealistic. It was their view that areas in which there was no catch data may not contain fish."

263 Morgan (n. 3) 188.

264 *Draft Summary Record Trilateral Management Meeting for SBT* (30 September 1991) (unpublished, copy held by author extracted from DPIE files)(hereinafter Draft 1991 Summary Record) 7–8.

265 Polacheck (n. 3) 292–293.

d *Decision-making in the* CCSBT *and the Duty of Cooperation*

The *Whaling* case is of no assistance for assessing the likelihood of success of the claims based on Articles 64 and 116–118 of UNCLOS, as their common element, the obligation of cooperation, is not mentioned in the ICRW.[266] In order to make out these claims, Australia and New Zealand would have to show that the impasse within the CCSBT could be laid only at Japan's door, not their own. This would not have been straightforward, given that under Article 7 of the 1993 Convention the CCSBT makes its decisions by consensus.[267]

The consensus rule is not without its problems. Compared with the (qualified) majority voting rule coupled with an objection procedure seen in the ICRW[268] and other fisheries commissions such as the IOTC,[269] it only superficially avoids the risk of deadlocks to which these commissions are exposed through the lodgement of objections, since the focus of disagreement and obstruction simply shifts to the making of the decision itself.

Thanks to each party's effective veto, in the CCSBT's early years it frequently took several postponements to unplanned extra sessions of meetings before an unchanged TAC and national allocations could be adopted, usually not until at least one of the members whose fishing season was about to start had adopted a self-imposed provisional quota in order to allow fishing to begin. Thus it was necessary to adjourn and convene a second session of the CCSBT's Third Meeting when the disagreement over experimental fishing resulted in no TAC being set at what subsequently became the first session.[270] Meanwhile Australia and New Zealand gave undertakings that, should there by the start of their next fishing seasons still be no agreed TAC and national allocations, they would act as though they continued to be bound by their previous catch limits.[271] Japan gave no comparable commitment, and when the meeting resumed in February 1997 it described the fact that Australia and New Zealand had com-

266 The ICJ did however twice mention the duty of States parties to the ICRW to cooperate with the IWC and its Scientific Committee: *Whaling in the Antarctic* (n. 1) 257 [83] and 297 [240].

267 For the text of art. 7, see (n. 148).

268 Text at (n. 10).

269 Agreement Establishing the Indian Ocean Tuna Commission (n. 238) arts V(3), IX (1)(2), permit the IOTC, by two-thirds majority of the Members present and voting, to adopt conservation and management measures binding on its Members. Such measures are subject to an objection procedure: art. IX(5) and (6).

270 CCSBT3(1) Report (n. 118) 22–23.

271 Ibid. 23.

menced fishing operations without a TAC as an "abnormal situation".[272] It is noteworthy that Japan stopped short of alleging breach of any obligation, as the proposition that no fishing at all would be permissible absent a TAC would have required Japan itself as well as Australia and New Zealand to refrain from further fishing. The better view is that, since no party in fact opposed continuation of fishing at the previous year's level, there can be no legal impediment to the course of action adopted by Australia and New Zealand.

Even so, whether this was a superior fisheries or legal policy outcome to having a TAC in force from which one party had opted out under an objection procedure must be doubted. Indeed, in some ways, at least from the perspective of Australia and New Zealand, it may well have been inferior, in that the exercise by Japan of its veto was obscured, substantially alleviating the degree of political discomfort that Japan would have felt as a consequence of its open use.[273]

A preferable model, it is suggested, is the decision-making procedure of the Auckland Convention,[274] which encourages in Article 16 the achievement of consensus but allows for decisions to be taken by a three-quarters majority. There is an objection procedure, but the grounds on which an aggrieved party may present an objection to a decision are severely limited: that it is contrary to one or more of UNCLOS, the UN Fish Stocks Agreement and the Auckland Convention itself, or discriminates unjustifiably in form or in fact against the objecting party.[275]

The 1993 Convention, however, was concluded several years too early for an equivalent rule to have been included within it, particularly in respect of the first limb, as this predates the first session of the conference at which the UN Fish Stocks Agreement was negotiated. Whether Japan would have

272 CCSBT, *Report of the Resumed Third Annual Meeting (Revised)* (18–22 February 1997) 2 <http://www.ccsbt.org/userfiles/file/docs_english/meetings/meeting_reports/ccsbt_03/report_of_ccsbt3_part2.pdf> accessed 5 August 2015.

273 See the remark by the Agent for Japan, Mr Togo, in the provisional measures phase of the dispute (Transcript for morning of 19 August (n. 103) 9), that use of the "veto", impliedly by Australia and New Zealand, was damaging the CCSBT, and the response by the Agent for Australia, Mr Campbell: *Southern Bluefin Tuna Cases (New Zealand v. Japan; Australia v. Japan)* (Requests for provisional measures) ITLOS doc ITLOS/PV.99/24/Rev.2 7–8 <https://www.itlos.org/fileadmin/itlos/documents/cases/case_no_3_4/VRE200899am.corr-rev.2.pdf> accessed 5 August 2015.

274 Convention on Conservation and Management of the High Seas Fishery Resources of the South Pacific Ocean (opened for signature 1 February 2010, entered into force 24 August 2012) UN registration no 50553 <https://treaties.un.org/doc/Publication/UNTS/No%20Volume/50553/Part/I-50553-080000280363a44.pdf> accessed 13 July 2015.

275 Ibid, art. 17(2).

consented to a reference to UNCLOS, not yet in force at that time, is equally open to doubt.[276] This would have left as the first limb the single ground of the decision being *ultra vires* the 1993 Convention alone. Under such a putative rule, it is suggested that Japan would have been compelled to argue that the TAC, by being set at too conservative (low) a level, was not capable of achieving the 1993 Convention's objective of "ensur[ing], through appropriate management, the conservation and optimum utilisation" of SBT. This in turn would have depended on the meaning, and relationship with each other, of the terms "conservation" and "optimum utilisation". These are considered in the next subsection below; for present purposes it suffices to observe that this would in effect have reversed the onus of proof that would have been borne by the applicants had the dispute in fact reached the merits stage—and, even without foreknowledge of that dispute, Japan would probably have perceived, and resisted, such a consequence. As for the second limb, Japan would have had to show that the national allocations adopted by the majority (Australia and New Zealand) discriminated unjustifiably against it. On the assumption that the national allocations making up the TAC would have been reduced *pari passu*, the arguments Japan would have employed can only be guessed at, but they could hardly have avoided being centred on the CCSBT failing to have sufficient regard to Japan's historical interests in fishing for SBT and its contribution to scientific research of the stock, as recently successfully argued by the Russian Federation in the only review to date of a decision under Article 17 of the Auckland Convention.[277] Such arguments, however, would have had to

276 In 'Summary Record' (Attachment F to *Southern Bluefin Tuna Trilateral Management Discussions Seventh Round* (ca 1988, unpublished, copy held by author extracted from DPIE files), hereinafter 7th Trilateral Management Meeting Report) 3, the fact that UNCLOS was not yet in force was the reason for Japan's objection to using the term "EEZ", despite what the ICJ had held in the *Gulf of Maine* case some years prior (n. 163 and accompanying text). Not long afterwards, however, Japan was among the States negotiating what became the Convention on the Conservation and Management of Pollock Resources in the Central Bering Sea (signed 16 June 1994, entered into force 8 December 1995) 34 *ILM* 67 which all accepted the high seas fisheries provisions (Part VII, Section 2) of UNCLOS as a statement of the relevant customary international law obligations: Stuart M. Kaye, *International Fisheries Management* (Kluwer Law International 2001) 322–323.

277 *In Proceedings Conducted by the Review Panel Established under Article 17 and Annex II of the Convention on the Conservation and Management of the High Seas Fishery Resources of the South Pacific Ocean with regard to the Objection by the Russian Federation to a Decision of the Commission of the South Pacific Regional Fisheries Management Organisation* (Findings and Recommendations of the Review Panel, 5 July 2013) <https://pcacases.com/web/view/33> accessed 24 July 2015; see also Andrew Serdy, 'Implementing Article 28

overcome Japan's acceptance in 1990 that in the long run its share of the catch as among the three original parties should fall to 47%.[278]

The net effect would thus have been that the fisheries management policy disagreement which gradually soured the atmosphere in the CCSBT from the mid-1990s before erupting in an open dispute in 1998 might well have been brought to the surface, and solved, much earlier. While there is a risk that a finding against Japan under a review procedure would have put unbearable

---

of the UN Fish Stocks Agreement: The First Review of a Conservation Measure in the South Pacific Regional Fisheries Management Organisation' (2016) 47 *Ocean Development & International Law* 1. Art. 8(4) of the 1993 Convention lists among the factors to be considered by the CCSBT in deciding Parties' national allocations:

. . .

(d) the interests of Parties whose vessels engage in fishing for southern bluefin tuna including those which have historically engaged in such fishing and those which have southern bluefin tuna fisheries under development;

(e) the contribution of each Party to conservation and enhancement of, and scientific research on, southern bluefin tuna;

. . .

278 This was the effect of a formula agreed in 1990 during the trilateral period for future increases in national allocations which, was confirmed and adopted by the CCSBT as its own at its First Meeting in 1994: CCSBT1 Report (n. 115) 2 and Annex 2. This would preclude any argument by Japan that it had borne a disproportionate share of the burden of conservation. If it wanted to rely on its contribution to the "enhancement" of the stock in art. 8(4)(e)—presumably a reference to the initial fishdown of the SBT stock from virgin biomass to generate reproductive surpluses (see infra, text between (nn. 286 and 289))—the answer to this would have lain in the observation that, while this does provide benefit for others, the overwhelming economic benefits are reaped by the pioneer fishing State itself through high catch rates: accord C.W. Armstrong, 'Co-operative Solutions in a Transboundary Fishery: The Russian-Norwegian Co-Management of the Arcto-Norwegian Cod Stock' (1994) 9 *Marine Resource Economics* 329, 337. Note that the factors do not include capacity as such (except possibly under (f) "any other factors which the Commission deems appropriate"), which would have been a profoundly anti-precautionary step. In 1997 Japan argued that the current allocation was inappropriate because it gave no weight to Members' historical catch records and fishing capacities and the social and economic dependence of their fishing industry on the SBT fisheries: CCSBT, *Report of the Fourth Annual Meeting, First Part* (8–13 September 1997) <http://www.ccsbt.org/userfiles/file/docs_english/meetings/meeting_reports/ccsbt_04/report_of_ccsbt4_part1.pdf> accessed 5 August 2015. It is submitted, however, that if the factors mentioned by Japan were given priority or even equal weighting with the others, it would act as a disincentive to reduce overcapacity. This would run counter to the inescapable imperative in a fishery for a depleted stock that capacity must be adjusted to the safe level of catch, not *vice versa*.

strain on its commitment to the CCSBT framework, Japan does not have a history of leaving bodies whose policy stance it finds uncongenial to its interests, as the whaling dispute itself attests. Rather, a procedure akin to that in the Auckland Convention, that ensures that TAC and national allocations can be adopted by a qualified majority and that minority objectors can escape being bound by them only by persuading a review panel that the decision is somehow legally defective, would seem better suited to prevent deadlocks that prevent a fisheries commission from fulfilling its basic task.

e *The Relationship between Conservation and Optimum Utilisation*

The question here is whether the goals of conservation and optimum utilisation mentioned in Article 64 of UNCLOS are of equal legal importance or priority, or whether one—conservation—is logically anterior to the other and must first be assured before weight can be given to optimum utilisation. According to Koers,

> [f]ull utilization and conservation are not opposing concepts. The central concern of conservation is to prevent the waste of the living resources of the sea by over-exploitation and to preserve their productivity for the future. This implies that any long term full utilization programme must take into account the demands of conservation, since the long range productivity of a stock is normally adversely affected by exploitation beyond its maximum biological limits.[279]

To the extent that coastal States dependent on local stocks systematically tend to favour conservation and distant-water fishing States with greater flexibility to seek fish elsewhere favour optimum utilisation, this issue may serve as a proxy for the question of the degree of subordination of the latter's interests to the former.[280] Where the balance should lie may depend on how the stock's biomass stands in relation to that which generates the maximum sustainable yield (MSY). For stocks significantly below that level, it is submitted that the primary obligation is to restore the biomass to that level, for the benefit not only of the present participants in the fishery but of all who might potentially wish to enter it. The role of optimum utilisation in these circumstances is at most to govern the speed of the rebuilding, i.e. to take economic factors into account, but not to the extent of stopping or reversing it.[281]

---

279 Koers (n. 103) 45.

280 See (n. 162) and accompanying text.

281 While it would have been in the applicants' interests to adopt this view, Japan is on record as insisting that the two goals of optimum utilisation and conservation contradicted,

17 Concluding Observations

While it would be a mistake to suppose that Australia and New Zealand would inevitably have prevailed in the merits phase of the *Southern Bluefin Tuna* arbitration had their case proceeded that far, there is enough in the foregoing analysis to support a conclusion that it was not doomed to failure simply because international courts and tribunals cannot cope with scientific arguments. The ICJ in the *Whaling* case showed itself adept at steering clear of deciding points on which scientific opinion was divided, while drawing damaging inferences on the discrepancies between the stated scientific aims of JARPA II and the low probability or in some instances impossibility of achieving these under that programme either as originally designed or as actually implemented. In the *Southern Bluefin Tuna* dispute there was ample scope for the Annex VII Tribunal to operate in and reason in the same way: the weaknesses of the ultimately propounded rationale for Japan's experimental fishing and the way its results would have been used were not such as to require expertise in population dynamics or any of the other fields making up fisheries science in order to unearth or understand. The flaws pointing to the conclusion that the measures implementing this fishing did not meet the standard laid down in UNCLOS Article 119, paragraph 1 of being based on the best scientific evidence available were ones that anyone capable of logical reasoning could spot in Japan's statements and the necessary consequences flowing from these.

On the other heads of claim in the various articles centred on the obligation to cooperate, however, it is somewhat less clear whether the Annex VII Tribunal would have avoided falling into the trap of thinking that the proper course is for States to adopt arbitrary average figures for particular parameters on which there is scientific disagreement, where the true position is much more likely to be closer to one of the contending camps' view than somewhere in the middle.[282] Cooperation in these circumstances must be aimed at

---

and must thus be balanced with, each other—see CCSBT, *Report of the Sixth Annual Meeting, First Part* (29, 30 November 1999) 4 [24] <http://www.ccsbt.org/userfiles/file/docs_english/meetings/meeting_reports/ccsbt_06/report_of_ccsbt6_Part1.pdf> accessed 14 August 2015, and, notably, its advocacy of optimum utilisation both in general and as a reason for raising the TAC by 3,000 tonnes in particular, in Attachment C ('Opening Statement by Japan at the Sixth Annual Meeting of CCSBT November 1999, Canberra') [4–5].

282 Accord Stephens (n. 3) 188:

"Encouraging the parties to a dispute to reach a compromise may well produce more harmonious relations but it will not necessarily lead to optimal environmental outcomes. In many cases it may serve to restore (or enhance) comity but only at the expense of the protection and preservation of the environment."

tackling the causes of disagreement, not their numerical expression. It would lead to unduly rigid constraint on the scientists as well as making it impossible to act on the best available evidence if, for example, their early view that the age at which SBT reach maturity was 8 were to be transformed into a proposition of law from which no delegation could thereafter depart without the agreement of the CCSBT's Scientific Committee, irrespective of new evidence controverting the previously agreed figure.[283] It is for reasons such as this that, while unilateral action such as that taken by Japan is not in itself proof of a breach of the duty to cooperate, it is certainly a strong grounding for provisional measures to be prescribed, and in retrospect it is from this phase of the dispute that the most benefit came.

Appendix: SBT Fishery Science

Maximum Sustainable Yield—The Surplus Production Model of Fisheries

Although most decisions on how fisheries are managed will expressly or impliedly be numerical in form, because fish stock assessment is an inexact science, the advice that scientists offer managers on the likely biological consequences of whatever management actions are under contemplation is invariably plagued by significant uncertainty. This centres on the balance between opposite risks: in the short term reducing catches can lead to an immediate economic loss, but failure to reduce them may deplete the stock, bringing about greater losses in the longer term. These risks can be represented mathematically in a number of different ways.[284]

Possibly the simplest of these is Schaefer's surplus production model. This rests on several assumptions that are reasonable approximations of how fisheries operate, even if they are unlikely to be encountered in the real world.[285]

283 In the provisional measures phase the use of age 12 by Australian and New Zealand scientists was criticised as "statistical sleight of hand" by Japan, followed by the contention that "there is no reason why this Tribunal should accept an age of maturity other than the one accepted by the Commission's Scientific Committee—and that is age 8": Transcript for morning of 19 August (n. 103) 23 (per Mr Greig, counsel for Japan).

284 For a useful and not excessively technical conspectus of the basic mathematics and statistics of fishery science, see the appendix in W. Harry Everhart and William D. Youngs, *Principles of Fishery Science* (2nd edn, Cornell University Press 1981) 294–336.

285 The account that follows is drawn from Milner B. Schaefer, 'Some Considerations of Population Dynamics and Economics in Relation to the Management of Commercial Marine Fisheries' (1957) 14 *Journal of the Fisheries Research Board of Canada* 669, 672–673.

Leaving aside perturbations from unpredictable environmental shocks, the model posits that fish populations tend in the long run to stay in dynamic balance, in other words, losses to the stock from all sources of mortality, both fishing and natural (predation, disease, senescence), are balanced by gains from a combination of the growth of individual fish that survive and an increase in their number.[286] As the rate of loss rises or falls, so does the rate of renewal, such that the balance is restored. The average natural rate of increase is a function of the size of the population, while the catch is also some function of both population size and fishing effort. The Schaefer model then assumes that: (a) the instantaneous (logarithmic) rate of fishing mortality is directly proportional to the chosen measure of fishing effort; and (b) the natural rate of increase of a fished stock at any instant is directly proportional to the difference between its biomass at that instant and the virgin biomass before fishing began, expressed as a long-term average. The rate of increase must be zero both when the population is zero and at virgin biomass, reaching a maximum at some intermediate value; this is the biomass generating the MSY (B_{msy}). A quadratic relationship is a "reasonably good first approximation"[287] (indeed elementary integral calculus dictates from the assumption of linearity in the rate of change in (b) above that the relationship must be a quadratic one)

The justification for these simplifications is that aiming for greater precision by taking into account large numbers of variables produces models that are too complex to be of use in improving understanding of the underlying natural phenomena: B.J. Rothschild and A. Suda, 'Population Dynamics of Tuna' in J.A. Gulland (ed), *Fish Population Dynamics* (John Wiley & Sons 1977) 309, 317.

286 Fish grow throughout their lives, but as they age the rate of growth becomes ever slower; according to the widely used von Bertalanffy growth equation, it reaches zero when the fish is infinitely old: see G.P. Kirkwood, 'Estimation of von Bertalanffy Growth Curve Parameter using both Length Increment and Age-Length Data' (1983) 40 *Canadian Journal of Fisheries and Aquatic Sciences* 1405, 1405, terming as "ubiquitous in the fisheries literature" equation (1) 1406, which is expressed in the form suitable for deriving a fish's age from its length, followed by a discussion at 1407–1409 of its application to SBT. When the number of fish competing for the same food sources declines, this will often manifest itself in empirical evidence such as an increase in the growth rate, as indeed has been observed for SBT since about 1980 in line with the inroads made into the stock: Attachment 6 to CCSBT-ESC Report 2004 (n. 51) 204. The phenomenon was first noted in *Report of the 9th Meeting of Australian, Japanese and New Zealand Scientists on Southern Bluefin Tuna* (17–22 September 1990) (hereinafter 9th Trilateral Scientific Meeting Report), in Trilateral Scientific Meeting Reports Compendium (n. 95) 75, 84.

287 Schaefer (n. 285) 673.

and under this model B_{msy} is normally estimated at around half of the virgin biomass.[288]

It will be noted that this model treats human harvesting of fish as merely an additional source of mortality by predation, whose effect is to raise the rate of renewal so as to restore the balance at some lower absolute level of biomass. A fishery can in theory continue indefinitely if it catches no more than the entire natural increase for the level of population, but this replacement yield can be low by comparison with MSY for stocks that are far from B_{msy}, in other words substantially underexploited or overexploited.

Of significance for SBT, the Schaefer model leaves out of consideration the age structure of the stock,[289] even though it is unavoidably altered by fishing and its effect on recruitment.[290] One reason for this is that the natural mortality of SBT varies with age; young fish have higher and old fish lower mortality.[291] Another is that the parental biomass may no longer reliably produce recruits once it has been reduced to less than half of its unfished level, as is the case

288 B_{msy} as half of virgin biomass is described as a "rule of thumb" in D.H. Cushing, *Science and the Fisheries* (Edward Arnold 1977) 32 and this too necessarily holds true if the equation describing the yield curve is quadratic: see the diagram in Andrew Serdy, *The New Entrants Problem in International Fisheries Law* (Cambridge University Press 2016) 18. It is accepted by some States (e.g. Canada in ICCAT, *Proceedings of the Thirteenth Regular Meeting of the Commission* (8–12 November 1993) in ICCAT, *Report for biennial period 1992–93* Part II (1993) 29, 39 [16a.6]). The Schaefer model can accommodate different assumptions, e.g. if a logarithmic rate of change is introduced into it in lieu of the linear one, the result is that B_{msy} occurs when the biomass is reduced to around 37% (the reciprocal of *e*, the base of natural logarithms) of the virgin biomass: William W. Fox Jr, 'An Exponential Surplus-Yield Model for Optimizing Exploited Fish Populations' (1970) 99 *Transactions of the American Fisheries Society* 80, 84.

289 The Beverton-Holt cohort model, which does take account of age structure, lends itself much less easily to economic modelling than the Schaefer surplus production model, since normally cohorts (fish of a stock spawned in the same year, also known as a year class) cannot be harvested individually: Gordon R. Munro and Anthony D. Scott, 'The Economics of Fisheries Management' in Allen V. Kneese and James L. Sweeney (eds), *Handbook of Natural Resource and Energy Economics*, vol. II (Elsevier 1985) 623, 625.

290 The average age of the stock at virgin biomass is reduced once fishing starts: W.E. Ricker, 'Stock and Recruitment' (1954) 11 *Journal of the Fisheries Research Board of Canada* 559, 583. This is a necessary consequence of new fish of age 0 replacing older fish whose removal from the ecosystem has made room for them, and in the initial "fish-down" phase of a fishery targeting a hitherto unfished stock, it is the accumulated older fish that tend to be caught first: Caton, McLoughlin and Williams (n. 52) 26.

291 Attachment 6 to CCSBT-ESC Report 2004 (n. 51) 204.

with SBT.[292] Among the chief risks for depleted stocks is that this state leaves them vulnerable to environmental variability that can cause a sharp decline in recruitment, which may in turn lead to collapse of the parental biomass. Although it is uncertain how low that biomass can be driven before it collapses, and not yet possible to predict the probability of collapse at a given level, the northern cod stock off Atlantic Canada is often cited as an example of a stock known to have collapsed in this way, and it is self-evident that, the lower the biomass and the longer it remains so, the higher must be the risk of such abrupt recruitment declines. An equation for the stock-recruitment relationship would be needed in order to quantify that risk, but information on this relationship seldom exists at lower stock sizes. Moreover, if at such sizes there is greater natural variability around the predicted recruitment than for larger stock sizes, or the normal processes determining recruitment break down altogether, past observations as to recruitment are no longer necessarily a reliable guide to the future.[293]

Virtual Population Analysis

Unlike the position in the International Commission for the Conservation of Atlantic Tunas, where MSY is specified as the management goal in its constitutive convention,[294] CCSBT scientific advice is given in terms not primarily of the MSY (though the Scientific Committee does now estimate it, as seen below[295]) but rather of the likelihood of rebuilding the parental biomass to the level of particular years under various assumed levels of catch. For this the technique of virtual population analysis[296] (VPA) is used. VPA projections can

---

292 Infra, text at (n. 298).

293 See Polacheck (n. 3) 292; Caton, McLoughlin and Williams (n. 52) 26; 13th Trilateral Scientific Meeting Report (n. 54) 5 [25].

294 International Convention for the Conservation of Atlantic Tunas (signed 14 May 1966, entered into force 21 March 1969) 673 UNTS 63. Art. IV(2)(b) speaks of the "maintenance of the populations of tuna and tuna-like fishes in the Convention area at levels which will permit the maximum sustainable catch and which will ensure the effective exploitation of these fishes in a manner consistent with this catch"; by contrast art. 3 of the 1993 Convention (n. 114) requires only the "conservation and optimum utilisation" of SBT.

295 Infra, text at (n. 328).

296 According to the definition of technical terms used by ICCAT (n. 77) 164–165, a VPA proceeds by analysing the catches from a given cohort over its life in the fishery, as follows: If, say, 10 fish were caught per year from the 1968 year class [cohort] for ten successive years from 1970 to 1979, then 100 fish would have been caught from that year class during its life in the fishery to that point. Since 10 fish were caught during 1979, then at least 10 fish must have been alive at the beginning of that year. Similarly there must have been

show, for a given set of assumptions, how the population size and structure would change under various catch combinations. They are, however, extremely sensitive to the information inputs used to derive them, particularly the form of the stock-recruitment relationship. Since the real relationship is unknown, any projection is simply a numerical representation of the assumptions used, both in the VPA and in the stock-recruitment relationship.[297]

For SBT, the drawback of VPA is that it can take several years to determine the impact of recent fishing on recruitment. In the interim, it may not be possible to detect a fall in recruitment sufficiently early to permit remedial reductions in catch that could prevent the fall becoming substantial, making

---

at least 20 fish alive at the beginning of 1978, at least 30 at the beginning of 1977 and at least 100 at the beginning of 1970. The VPA calculates the number of fish that must have been alive if some fish also died from causes other than fishing. If the instantaneous natural mortality rate was known in addition to the 10 fish caught per year in the fishery—and normally this is known within a fairly small range—then the VPA calculates the number that must have been alive each year to produce a catch of 10 fish per year in addition to those that died from natural causes. If the fishing mortality rate for the last year for which data are available is known, then the exact abundance of the year class can be determined in each year if the catches are known with certainty. Thus VPA eventually allows an estimation of the number of recruits produced each year. Until the development of the technique of counting rings on the otolith (see J. Thorogood, 'Age and Growth Rate Determination of Southern Bluefin Tuna, *Thunnus maccoyii*, using otolith banding' (1987) 30 *Journal of Fish Biology* 7), the age of individual fish was formerly difficult to assess directly, becoming progressively more so as size increases. Instead the length composition of the catch was determined by sampling, from which it was possible to estimate the numbers of fish caught at each length of the entire catch. These lengths were then converted to ages through an accepted age-length relationship. See Kirkwood (n. 286); Caton, McLoughlin and Williams (n. 52) 25. The otolith is a calcareous concretion in the inner ear of fishes, laid down in concentric layers which in most species have slight but observable colour or density variations. The cause of these variations is unknown, but probably related to the passage of the seasons either directly or, through cyclical changes in diet, indirectly: Everhart and Youngs (n. 284) 63. For some problems and shortcomings of cohort analysis see Tony J. Pitcher and Paul J.B. Hart, *Fisheries Ecology* (Croom Helm 1982) 374–377.

297 Caton, McLoughlin and Williams (n. 52) 26. In 1994 the scientists reported that the relationship between parental biomass and recruitment at the recent low levels of parental biomass was unknown for SBT, with all VPAs used in the assessment showing a parental biomass of about 50% of the lowest level for which reasonably precise recruitment estimates were available. Some compensation (i.e. increased recruitment per unit of parental biomass) was observed in the estimated recruitment between 1965 and 1989, for which the Japanese and Australian VPAs indicated very similar trends: 13th Trilateral Scientific Meeting Report (n. 54) 2 [8–9], 4 [15].

it all the harder to achieve stock recovery. The VPA results were open to various interpretations owing to the uncertainty in input data, but as the models for SBT were progressively refined over the 1980s, a reasonable understanding developed of the extent to which this uncertainty influenced the results, and the downward trend in parental biomass was consistent notwithstanding the uncertainties. VPAs presented at the 1989 scientific meeting showed that the parental biomass in 1988 was at worst 8% and at best 25% of that in 1960, which itself was below the unfished level by an unknown amount, and confirmed previous predictions that a further decline for at least another year or two was inevitable.[298]

In the 1990s the accuracy of VPAs for SBT came to depend on the rate of population decline and the correctness of the starting value of the fishing mortality rate, which required additional information in the form of indices of abundance or fishing mortality rates (commonly referred to as tuning indices). Indices of age-specific catch and effort data from the Japanese longline fishery were the primary source of information for tuning, but analysis of these data was limited by their aggregated nature, all catch and effort within 5° squares of latitude and longitude being pooled month by month. Worse still, the size data used to estimate the age distribution of the catches were pooled quarter by quarter in blocks of 5° of latitude and 10° of longitude, not all of which, even at this level of aggregation, were sampled. In addition, data related to changes in fleet efficiency were lacking, making it difficult to distinguish the effects of changes in abundance from changes in efficiency in the observed catch rates. Nevertheless, the standard statistical analyses performed on these aggregated data indicated significant year-to-year variation in relative catch rates or densities among areas and seasons.[299]

Scientific Advice to Managers

The report of the scientists' 1988 meeting summarises the history of their joint endeavours.[300] Although the scientists at previous meetings had stressed that SBT parental biomass levels below those in 1980 were likely to cause a fall in recruitment, the evidence had been insufficient to conclude that this was occurring. Unsure of their ability to detect a decline early enough to advise managers on actions to reverse it, they had recommended substantial reductions in the global catches of SBT. In 1986, as a consequence of the accumulating catch and effort data and improvements in stock assessment methods, the

298 Caton, McLoughlin and Williams (n. 52) 24. See also (n. 123).

299 Polacheck (n. 3) 285–287.

300 7th Trilateral Scientific Meeting Report (n. 103) 42–43.

scientists became more concerned about a possible fall in recruitment. In 1987 the scientists recommended *inter alia* that

> 1. Managers should recognise that there is a risk associated with maintaining the current catch limits. If catches were reduced, the risk would decrease.
> 2. Governments of the three countries take immediate steps to ensure that future major reductions in catch, if necessary, can be implemented quickly and effectively.[301]

By 1988, all VPA results were indicating that by 1979 the SBT biomass had fallen to a dangerously low proportion of its unexploited level, with recruitment having begun to decline before 1980 under some plausible assumptions, or fluctuating without apparent trend under others. For the period 1979–1987 the VPAs again uniformly showed that the parental biomass had suffered a considerable additional decline. Largely due to the high Australian catches of the early 1980s, under most plausible combinations of assumptions, at the then current level of catches, both parental biomass and recruitment were predicted to decline even further over the next few years, leading to a collapse of the population.

Though they described the need to decrease catches as "clear cut", the scientists declined to specify the extent of the required reduction. Citing the uncertainty about the dynamics of the SBT population, they stated that only with a complete cessation of catching could they be confident of the stock's recovery—though even with zero catch there existed biological mechanisms that meant recovery could not be guaranteed. Correctly anticipating that managers would prefer not to reduce catches to zero, the scientists unanimously advised that immediate reductions of at least half of current catches should be made to all sectors of the fishery, and not reversed until significant improvement to the SBT stock's status could be demonstrated.[302]

---

301 *Report of the sixth meeting of Australian, Japanese and New Zealand scientists on southern bluefin tuna* (17–21 August 1987) in Trilateral Scientific Meeting Reports Compendium (n. 95) 35, 37.

302 7th Trilateral Scientific Meeting Report (n. 103) 43–44. Even this was accepted only with difficulty by Japan in the 1988 management meeting, where it stated (7th Trilateral Management Meeting Report (n. 276) 6) that it "could not proceed to determine management measures without exploring stock projections" which it expected to be more substantive, suggesting that the scientists be given more time "to refine analyses and come up with more definite recommendations". It listed a string of apparently positive indicators "which prove that the 1982 SBT cohort might not have been as damaged as is thought". It suggested that "other factors associated with oceanography and climate may have influ-

Though there was no dissent from this diagnosis, and major catch reductions were duly imposed in 1988 and 1989, from that point it was the prospects for the stock's recovery that divided the scientists.[303] At the 1989 scientific meeting no agreement could be reached, with different Australian and Japanese views presented in the report. The Australian argument was that the exclusion from the Japanese scientists' analysis of historic and future Indonesian, Korean, Taiwanese and New Zealand catch, and of SBT bycatch in other fisheries, had the effect of underestimating the extent of the decline in parental biomass to 1988 and afterwards. Moreover, their assumption that recruitment did not decline on average between 1960 and 1975 was optimistic and had led them to overestimate the likely stock recovery.[304] The New Zealand scientist present concluded that there was a risk of driving the parental biomass below a critical mass at which it could sustain itself.[305]

---

enced the stock situation and enquired if the scientists had considered all those factors." Clutching at straws to forestall catch reduction, Japan in effect even disowned its own scientists: it "questioned the objectivity of the scientific report in choosing particular years so as to present more pessimistic stock forecasts." See also October 1989 Draft Summary Record (n. 123) 5, where Japan believed that catches up to 1980 were at or below the MSY for the SBT fishery (implicitly blaming the subsequent decline on the large Australian catches of small fish in 1982).

303 It may not be coincidental that, as pointed out by Polacheck (n. 3) 285, while the earlier catch limits reduced catches and fishing mortality rates from the surface fisheries, it was not until the 1989 fishing year that the catch limits became restrictive for the Japanese longline fleet. That is, until 1989 the Japanese fishery had not been able to reach its catch limit.

304 *Report of the Eighth Meeting of Australian, Japanese and New Zealand Scientists on Southern Bluefin Tuna* (4–10 September 1989) in Trilateral Scientific Meeting Reports Compendium (n. 95) 47, 53 and 61.

305 Ibid. 60. This would be contingent on the SBT stock being subject at very low levels to what is known as critical depensation, which was not asserted. (A stock is said to be subject to critical depensation if there is a biomass level below which it cannot sustain itself, so that the stock inevitably dies out—on depensation models see Colin W. Clark, *Mathematical Bioeconomics: The Optimal Management of Renewable Resources* (2nd edn, John Wiley & Sons 1990) 17. In other words the growth curve exhibits the property that, for very small values of the population, the growth rate is negative, meaning that the stock can be condemned to extinction by reducing it below the level at which the growth becomes negative, since beyond that point of minimum viability not even complete cessation of fishing will prevent its extinction. An example of a natural depensatory mechanism given by Ricker (n. 290) 602, is predation on very vulnerable fry by fish that eat their fill of them and move on, however many or few the fry may be. Note that the Schaefer model, (nn. 285–289) and accompanying text, disregards depensation: Clark (this n.) 50.

Little changed in the following years. In both 1990 and 1991 Japanese scientists took the view that the probability was very high that the stock would increase under current catch levels and even under slightly higher ones. The Australian and New Zealand scientists' view was that the stock might well already be in the process of recruitment collapse. Because the information was ambiguous, all were prepared to recommend that no increase in catch should be considered until there was clear scientific evidence of recovery in the parental stock.[306] On this point, however, while the Japanese scientists believed that the most reasonable range of assumptions led to projections which showed stock increases, their Australian and New Zealand counterparts found a broader range of assumptions plausible; some of the projections these yielded resulted in declines.[307]

---

306 9th Trilateral Scientific Meeting Report (n. 286) 91; *Report of the Tenth Meeting of Australian, Japanese and New Zealand Scientists on Southern Bluefin Tuna* (23–29 September 1991) (hereinafter 10th Trilateral Scientific Meeting Report), in Trilateral Scientific Meeting Reports Compendium (n. 95) 99, 102 [10] (Japan) and [11] (Australia and New Zealand)). The latter meeting nonetheless saw agreement (ibid. 99 [3]) that there had been:

"a continuous decline from 1980 to 1989 in the parental stock; a sharp decline from 1980 until 1986 or 1987, and thereafter a slight increase, in the pre-adult stock; an increase in small fish availability in many fishing grounds, an increase in CPUE and the reappearance of middle-sized fish on many fishing grounds, indications that escapement from the Australian surface fishery is increasing, but uncertainty whether present recruitment will guarantee recovery of the parental stock."

Similar recommendations against increases in catch were made in 1992 and 1993: 11th Trilateral Scientific Meeting Report (n. 120) 5 [19] and 12th Trilateral Scientific Meeting Report (n. 112) 5–6 [22] respectively.

307 10th Trilateral Scientific Meeting Report (n. 306) 101 [6]. In the management meeting New Zealand said that its 1991 catch of only 41% of its limit was "primarily thought to be a result of the poor state of the resource": Draft 1991 Summary Record (n. 264) 2. By contrast, the picture provided by Japan to ICCAT, of which Australia and New Zealand were not members, was based only on its own optimistic assessments: Panel 3 heard that the trilateral scientific meeting determined that projections from current catch levels would see the parent stock "reach its lowest level in 1990 or 1991", after which "[m]any projections predict recovery, even under increased catches of up to 20,000 [tonnes], with the parent stock attaining 1980 biomass levels by 2010", with the possible revision of the catch limits maintained since 1989 "currently being considered as a result of this year's scientific discussions": ICCAT, *Report of the Meetings of Panels 1 to 4* (Annex 6 to ICCAT, *Proceedings of the Seventh Special Meeting of the Commission* (12–16 November 1990) (hereinafter ICCATSM7 Report)), in ICCAT, *Report for biennial period* 1990–91 Part I (1990) 60, 71–72. A somewhat fuller story was told to the Standing Committee on Research and Statistics, with "socio-economic factors" (see also text at n. 109) given as an additional reason for revising the catch limits, along with the scientists' recommendation "that there be no

One of the factors in the long stasis initially was the contradictory signals from successive years of data. The Australian and New Zealand scientists recommended in 1991 that current catch levels should not continue beyond 1992 unless there was sufficient evidence to refute a strong fall in recruitment since the mid-1980s.[308] That evidence initially came in 1992, when new information showed that recruitment collapse had not occurred up to 1988, although concern remained about continued decline and the possibility of a future sudden fall while the parental biomass remained at its very low level.[309] The report stated that

> [d]ecreasing catch levels will increase the speed and possibility of recovery to the 1980 level of parental biomass. Maintaining the present catch level is expected to result in the low parental biomass seen in the 1980s continuing for many more years, so that the risk of an abrupt and unpredictable decline in recruitment would remain at about its present high level. A decrease in catch level is preferable from the biological viewpoint.[310]

Despite this, in view of the earlier catch reductions by the three States, the meeting stopped short of recommending a change in trilateral catch limits, merely encouraging changes to fishing practices to decrease targeting of SBT less than 15 kg and greater than 100 kg, though it called for "urgent and decisive management action" to be taken in the event of significant further declines.

---

increase in present catch levels until there is clear scientific evidence of a recovery in the parental stock": ICCAT, *Report of the Standing Committee on Research and Statistics* (*SCRS*) (5–9 November 1990) (Annex 10 to ICCATSM7 Report) ibid. 137, 193.

308 10th Trilateral Scientific Meeting Report (n. 306) 102 [11]. Again ICCAT's Panel 3 was told only one side of the story: that the lowest level should now be reached "in 1991 or soon thereafter": ICCAT, *Reports of the Meetings of Panels 1–4* (Annex 6 to ICCAT, *Proceedings of the Twelfth Regular Meeting of the Commission* (11–15 November 1991) (hereinafter ICCAT12 Report)), in ICCAT, *Report for biennial period* 1990–91 Part II (1991), 51, 57, while the Standing Committee on Research and Statistics was informed (ICCAT, *Report of the Standing Committee on Research and Statistics* (SCRS) (4–8 November 1991) (Annex 16 to ICCAT12 Report) ibid. 97, 135) that "most of the VPA projections show parent stock recovery. It suggests that the current regulations are effective for long-term southern bluefin tuna stock recovery."

309 11th Trilateral Scientific Meeting Report (n. 120) 3 [7].

310 Ibid. 5 [17]. The point in the last sentence was repeated the following year: 12th Trilateral Scientific Meeting Report (n. 112) 4 [16].

Instead, the meeting recommended that a strategy be developed for management action "to improve the chances of rebuilding the parental biomass to 1980 levels."[311]

At the 1993 scientific meeting, the advent of direct ageing techniques for SBT[312] led to significant revisions of the growth assumptions.[313] It thus became evident (with "very serious and detrimental" consequences for prospects of stock rebuilding) that several recent cohorts had not contributed to stock rebuilding.[314] The reinterpretation meant that some positive signs previously reported were no longer as apparent: some year-classes that had passed through the fishery since the large reduction in surface fishery catches were well represented in the longline fishery as young fish, but had not resulted in rebuilding of older age groups in subsequent years. This was thought to be due to the reduced size of these year-classes and the large catches taken from them. VPA results showed no convincing evidence of recent increases in the number of young mature fish, with parental biomass still decreasing steadily to at least 1992, rather than declining slowly or having become stable as previously thought. The combination of recent increases in the longline catch of juvenile SBT and the continuing, if diminished, Australian surface fishery limited the potential for recovery of the parent stock.[315]

---

311 11th Trilateral Scientific Meeting Report (n. 120) 5 [17–18]. This was reflected in the report of the following year's meeting, when the recommendation against raising catches until the parental biomass returned to the 1980 level (n. 306) was newly qualified by the words "unless this is part of an agreed stock rebuilding strategy that can be shown to have a high probability of returning the stock to biologically safe levels."

312 (n. 296).

313 12th Trilateral Scientific Meeting Report (n. 112) 1 [2] and 3 [7]. Specifically, the growth rate in the first year of life was faster than assumed in previous stock assessments; the overall growth rate of SBT had increased between the 1960s and 1980s; and the average size at maturity, about 145 cm, was larger than assumed in previous analyses (about 130 cm). The change in growth was substantial. For example, fish 100 cm long, which in the past had been taken as being about 5 years old, were now interpreted as being aged about 3 if caught since 1980. The new information thus exposed a misinterpretation of the increased catch of middle-sized fish as a general rebuilding of the juvenile age classes; in fact these were faster-growing young fish, and the catch rate for fish actually of the ages concerned showed no significant increase.

314 12th Trilateral Scientific Meeting Report (n. 112) 2 [7].

315 Ibid. 3–4 [8–12]. The growth revision also affected the recruitment projections, with the decline during the early 1980s to about 50% of the 1980 level by 1985 being "faster and greater than expected from standard stock-recruitment relationships, raising serious doubts about the ability of the population to recover under current catch levels." Projections based on the improved information about growth indicated that the pop-

After the gloom of the 1993 stock assessment, that of 1994 was again optimistic, with the scientists reporting that, although the parental biomass continued its decline to a historically low level in 1993, the VPA results indicated that the quota reductions were having an effect, with fishing mortality rates in the early 1990s being less than those in the 1980s.[316] The surface fishery catch reductions during the 1980s and increased recruitment in the late 1980s had resulted in an increase in the CPUE of the juvenile stock (fish of ages 3 to 7) since 1986. There was clear sequential rebuilding in CPUE of juvenile year classes, reaching 6-year-olds in 1993.[317] As the recent VPA recruitment estimates were all "well above the minimum level required to rebuild the stock given current catch levels", stock recovery would be "assured" if these recruitments and catches were maintained during the 1990s. If the stock-recruitment relationship were highly compensatory, then these recruitment levels could be expected to be maintained on average, but not if it were only moderately compensatory.[318]

---

ulation was expected to continue declining under constant catches, with only a small probability that the parental biomass would increase from the current low levels by 2010. Projections using the previous growth assumptions had indicated a probable slow recovery, with the parental biomass expected to remain below its 1980 level for about 15 years: ibid. 4 [13–14].

316 13th Trilateral Scientific Meeting Report (n. 54) 8–9 (Appendix 1 ('Status of the stock and fishery indicators') [4]).

317 Ibid. 9 (Appendix 1, [5–6]. By sequential rebuilding is meant the fact that the large increase in 3-year-olds in 1990 could be followed in successive years of catch rate data through to SBT aged 4 in 1991, aged 5 in 1992 and aged 6 in 1993, though its magnitude depended on the interpretation of the 1993 data. Sounding a cautionary note, the scientists added that the simultaneous increase in 1993 in the catch rate of all ages of SBT from 3 to 10 suggested a general increase in catchability superimposed on the sequential rebuilding of age groups, i.e. only an unknown part of the increased catches reflected an actual rise in abundance.

318 Ibid. 5 [26]. The Japanese analysis used two stock-recruitment relationships (one highly compensatory, the other moderately so) within one particular VPA and one interpretation of catch rates, while the Australian analysis used a range of VPAs and catch rate interpretations with a single moderately compensatory stock-recruitment relationship; ibid. 4 [17]. Under a highly compensatory relationship high recruitment is maintained, on average, even at very low parental biomass levels. A moderately compensatory relationship predicts that average recruitment decreases as parental biomass decreases. While the Australian scientists regarded the highly compensatory relationship as an unlikely description of average recruitment for SBT, their Japanese counterparts considered both highly and moderately compensatory relationships equally likely. With a moderately compensatory relationship the parental biomass was expected to remain below the 1980 level for many more years, while under the highly compensatory hypothesis recovery would be faster. See ibid. [23–25].

In most of the VPAs an increase in parental biomass was calculated for 1994. This was due to the continued catches at relatively constant rates of fish aged 12 and older, confounding the assumptions used in previous years, which had projected that by 1993 there would be insufficient fish of these ages to sustain the catches actually observed.[319] Given the estimated age structure of the current population, all projections indicated that the parental biomass would increase from the 1993 level over the next few years if the age composition of the longline catch remained similar to the past.[320]

In the years leading up to the 1998–2001 dispute over Japan's unilateral experimental fishing programme, the SBT stock assessments continued to be characterised by large uncertainties in the input data and the appropriate biological parameters, which were ultimately to culminate in the dispute. These included recent catch levels, natural mortality rates, age of maturity, different models for estimating the size of the older age classes of the population and the CPUE indices, which remained the primary tuning indices for the VPA. In the 1998 assessments, varying assumptions about the mean age of maturity made the trend in the parental biomass after 1994 uncertain, but all results for ages of maturity greater than age 8 indicated that it had continued to decline. In addition, recruitment was estimated to have markedly declined from the late 1960s to about a third of the 1960 level. The 1998 data also indicated that there was no increase in recruitment between 1988 and 1992 (the most recent cohort for which a direct estimate was available because of the VPA time lag), but CPUE and other indices for the 1992 to 1998 cohorts suggested that recruitment had remained low.[321]

---

319 13th Trilateral Scientific Meeting Report (n. 54) 3 [10–11]. Even so, almost all the VPAs indicated that parental biomass had continued to decline, albeit at a slowing rate, up to 1993. Though the Japanese and Australian VPAs differed in their estimates of the actual amount of parental biomass, they gave similar trends of change in that biomass. Its absolute scale, while difficult to estimate from VPA, did not greatly influence the trend: ibid. 2–3 [9]. All 1994 VPAs showed that the recruitment each year from 1986 to 1989 was higher than the low level of 1985, though with higher uncertainty surrounding the more recent of these estimates. The interpretations ranged from a pattern of substantial increase to 1989 to a moderate increase to 1987 followed by smaller declines in each of 1988 and 1989 to a level slightly above that in 1985: ibid. 3 [13] and 10 (Appendix 1 [8]).

320 Ibid. 4 [18] Subsequent change would depend on the level of recruitment of the early 1990s and the catch levels in the next few years, most projections indicating a levelling off or slight decline in five or six years. This dip would last for about five years and whether it would be followed by increases or decreases depended on the year-class strength of fish spawned in the 1990s and the catch level: ibid. [19].

321 Polacheck (n. 3) 285.

Because of inconsistencies among the temporal trends in standardised CPUE indices for the different age-classes, a range of possible interpretations, hypotheses and model structures were considered, and weighted averages of the results were used to provide managers with advice on the current and projected stock status. The stock assessments undertaken by national scientific delegations within the CCSBT relied on different sets of hypotheses, model structures and weightings. Even so, they yielded robust conclusions about the status of the stock relative to historical levels, all indicating that it was highly depleted. The Scientific Committee repeatedly concluded that "the continued low abundance of the SBT parental biomass is cause for serious biological concern".[322]

Stock projections under current catch levels indicated a wide range of possibilities, from rapid recovery to further substantial decline, depending upon which specific hypotheses were used to model uncertainties. The differences depended on the uncertainties considered, the weights assigned to different hypotheses and models, and the criteria (or lack of them) used for rejecting model prediction for lack of fit with the observed data. Thus, in 1998, the estimates by Australian and New Zealand scientists indicated a low probability of recovery of 14% or less, while those of the Japanese scientists were in a much higher range: 76% to 87%. The Australian and New Zealand estimates also put at greater than 50% the probability that the parental biomass would continue to decline. Retrospective analyses of the 1998 stock projections indicated that the estimates of probability of recovery decreased when updated with an additional year's catch and effort data. This was consistent with the pattern of past years, when the projections on which the Scientific Committee had based its advice were repeatedly shown to be overly optimistic in the light of subsequent stock assessment results.[323]

The CCSBT was unable to develop an agreed approach for dealing with the underlying disparity and uncertainty in the VPA projection results. The differences in the national delegations' estimates of the probability of recovery were attributable to the different weights—which in some cases were zero expressly or by implication—they assigned to the different hypotheses

322 Ibid. The phrase is used in the 1995 and 1998 reports of the Scientific Committee: the reports are unpublished but were quoted by the Agent for Australia, Mr Campbell, before ITLOS: Transcript for morning of 18 August (n. 257) 20 and 21 respectively. The same phrase had been used in the trilateral meeting reports to management in 1992 to 1994: 11th Trilateral Scientific Meeting Report (n. 120) 3 [6], ("serious" preceded by "very"); 12th Trilateral Scientific Meeting Report (n. 112) 2 [6]; 13th Trilateral Scientific Meeting Report (n. 54) 2 [7].

323 Polacheck (n. 3) 285–286.

and models used to quantify the underlying uncertainty in the stock assessments. The major identified factors changed somewhat over time as new data became available and as the assessment models evolved. In the late 1990s these included the form of the stock-recruitment relationship, natural mortality rates, the model for estimation of the numbers of older fish, interpretation of CPUE as indices of abundance and lack of fit of models to the observed data.[324]

After the end of the dispute the stock assessment reports again became unanimous, though on occasion they referred to differences of emphasis among the delegations.[325] At first their advice was that since the full surplus was being taken, the stock had an equal chance of growing or declining;[326] but from 2004 they again turned pessimistic because of successive years of low recruitment.[327]

The most recent stock assessment, that of 2014, found that the stock remained at a very low state well below B_{msy}, but with some improvement observed since the previous (2011) assessment in the form of a modest recovery in the biomass of fish aged 10 or more, now estimated at 7% of that when fishing began, up from 5% in 2011; the fishing mortality rate too was below the level associated with MSY.[328] It is thus clear that recovery of the SBT stock to B_{msy} as called for by Articles 61 and 119 of UNCLOS remains some way off under any management strategy involving continued fishing.

---

324 Ibid. 286. By comparison, in 1990 the factors highlighted had also included an unknown stock-recruitment relationship and uncertainty in both the natural mortality rate and the relationship between CPUE and abundance, but at that time the others were catches not fully accounted for, uncertainty in the age composition of the catch and the time lag in estimation of recruitment: 9th Trilateral Scientific Meeting Report (n. 286) 75.

325 The first few are CCSBT-SAG2 Report (n. 54); CCSBT-SAG3 Report (n. 54); CCSBT, *Report of the Fourth Meeting of the Stock Assessment Group* (25–29 August 2003) <http://www.ccsbt.org/userfiles/file/docs_english/meetings/meeting_reports/ccsbt_10/report_of_sag4.pdf> accessed 14 August 2015; CCSBT, *Report of the Fifth Meeting of the Stock Assessment Group* (6–11 September 2004) (hereinafter CCSBT-SAG5 Report), <https://www.ccsbt.org/sites/ccsbt.org/files/userfiles/file/docs_english/meetings/meeting_reports/ccsbt_11/report_of_sag5.pdf> accessed 25 August 2016.

326 As summarised by the Chair of the Scientific Committee: CCSBT8 Report (n. 230), 6–7 [43].

327 See CCSBT-SAG5 Report (n. 325) 3–4 [21] and 8 [48].

328 CCSBT-ESC Report 2015 (n. 70) [173] and Table 5.

References

Agreement Establishing the Indian Ocean Tuna Commission (Rome, 25 November 1993) 1927 UNTS 329.

Agreement for the Establishment of the Indo-Pacific Fisheries Council (Baguio, 26 February 1948) 120 UNTS 59.

Agreement for the Implementation of the Provisions of the United Nations Convention on the Law of the Sea of 10 December 1982 relating to the Conservation and Management of Straddling Fish Stocks and Highly Migratory Fish Stocks (New York, 4 August 1995) 2167 UNTS 3.

Agreement on Fisheries between the Government of Australia and the Government of Japan (Canberra, 17 October 1979) 1217 UNTS 3.

Agreement on Fisheries between the Government of New Zealand and the Government of Japan (Wellington, 1 September 1978) 1167 UNTS 441.

Armstrong, C.W., 'Co-operative Solutions in a Transboundary Fishery: The Russian-Norwegian Co-Management of the Arcto-Norwegian Cod Stock' (1994) 9 *Marine Resource Economics* 329.

Boyle, A., 'Problems of Compulsory Jurisdiction and the Settlement of Disputes Relating to Straddling Fish Stocks' in O.S. Stokke (ed), *Governing High Seas Fisheries: The Interplay of Global and Regional Regimes* (Oxford University Press 2001) 91.

Boyle, A., 'The Southern Bluefin Tuna Arbitration' (2001) 50 *International and Comparative Law Quarterly* 447.

Bureau of Rural Resources, *Reports of the Trilateral Scientific Discussions among Australia, Japan and New Zealand on Southern Bluefin Tuna 1982–1991* (Working Paper No. WP/10/92; Bureau of Rural Resources, 1992).

Burke, W.T., *The New International Law of Fisheries: UNCLOS 1982 and Beyond* (Clarendon Press 1994).

Butterworth, D.S., J.N. Ianelli and R. Hilborn, 'A Statistical Model for Stock Assessment of Southern Bluefin Tuna with Temporal Changes in Selectivity' (2003) 25 *African Journal of Marine Science* 331.

Butterworth, D.S., and A.J. Penney, 'Allocation in High Seas Fisheries: Avoiding Meltdown' in A.I.L. Payne, C.M. O'Brien and S.I. Rogers (eds), *Management of Shared Fish Stocks* (Blackwell Publishing Ltd 2003) 165.

Campbell, H., S.F. Herrick Jr and D. Squires, 'The Role of Research in Fisheries Management: The Conservation of Dolphins in the Eastern Tropical Pacific and the Exploitation of Southern Bluefin Tuna in the Southern Ocean' (2000) 31 *Ocean Development & International Law* 347.

Caton, A., K. McLoughlin and M.J. Williams, *Southern bluefin tuna: scientific background to the debate* (Australian Government Publishing Service 1990).

Churchill, R.R., and A.V. Lowe, *The Law of the Sea* (3rd edn, Manchester University Press 1999).

Clark, C.W., *Mathematical Bioeconomics: The Optimal Management of Renewable Resources* (2nd edn, John Wiley & Sons 1990).

Collette, B.B., and C.E. Nauen, *FAO Species Catalogue Vol. 2: Scombrids of the World: An Annotated and Illustrated Catalogue of Tunas, Mackerels, Bonitos and Related Species Known to Date* (FAO Fisheries Synopsis No 125 vol 2; FAO 1983).

Colson, D.A. and P. Hoyle, 'Satisfying the Procedural Prerequisites to the Compulsory Dispute Settlement Mechanisms of the 1982 Law of the Sea Convention: Did the Southern Bluefin Tuna Tribunal Get It Right?' (2003) 34 *Ocean Development & International Law* 59.

Convention for the Conservation of Southern Bluefin Tuna (Canberra, 10 May 1993), 1819 UNTS 359.

Convention on Conservation and Management of the High Seas Fishery Resources of the South Pacific Ocean (Auckland, 14 November 2009), UN registration no 50553 <https://treaties.un.org/doc/Publication/UNTS/No%20Volume/50553/Part/I-50553-080000028036 3a44.pdf> accessed 13 July 2015.

Convention on the Conservation and Management of Pollock Resources in the Central Bering Sea (Washington DC, 16 June 1994) 34 *International Legal Materials* 67.

Cushing, D.H., *Science and the Fisheries* (Edward Arnold 1977).

de Yturriaga, J.A., *The International Regime of Fisheries: From UNCLOS 1982 to the Presential Sea* (Martinus Nijhoff Publishers 1997).

Declaration on the Conduct of Parties in the South China Sea (Phnom Penh, 4 November 2002) <http://www.asean.org/?static_post=declaration-on-the-conduct-of-parties-in-the-south-china-sea-3&category_id=32> accessed 2 April 2016.

Delimitation of the Maritime Boundary in the Bay of Bengal (Bangladesh/Myanmar) (Judgment) [2012] ITLOS Rep 4.

Delimitation of the Maritime Boundary in the Gulf of Maine Area (Canada/United States of America) (Judgment) ICJ Rep 1984 246.

Everhart, W.H., and W.D. Youngs, *Principles of Fishery Science* (2nd edn, Cornell University Press 1981).

Foster, C.E., 'The "Real Dispute" in the Southern Bluefin Tuna Case: a Scientific Dispute?' (2001) 16 *International Journal of Marine and Coastal Law* 571.

Fox, W.W. Jr, 'An Exponential Surplus-Yield Model for Optimizing Exploited Fish Populations' (1970) 99 *Transactions of the American Fisheries Society* 80.

Fu, C., R. Mohn and L.P. Fanning, 'Why the Atlantic cod (*Gadus morhua*) stock off eastern Nova Scotia has not recovered' (2001) 58 *Canadian Journal of Fisheries and Aquatic Sciences* 1613.

Fujinami, N., 'Development of Japan's Tuna Fisheries' in D.J. Doulman (ed), *Tuna Issues and Perspectives in the Pacific Islands Region* (Pacific Islands Development Program 1987) 57.

In Proceedings Conducted by the Review Panel Established under Article 17 and Annex II of the Convention on the Conservation and Management of the High Seas Fishery Resources of the South Pacific Ocean with regard to the Objection by the Russian Federation to a Decision of the Commission of the South Pacific Regional Fisheries Management Organisation (Findings and Recommendations of the Review Panel, 5 July 2013) <https://pcacases.com/web/view/33> accessed 24 July 2015.

In the Matter of an Arbitration before an Arbitral Tribunal Constituted under Annex VII to the 1982 United Nations Convention on the Law of the Sea (Republic of the Philippines v. People's Republic of China) (Award on Jurisdiction and Admissibility, Order of 29 October 2015) PCA Case No. 2013–19, <http://www.pcacases.com/web/send Attach/1506> accessed 2 April 2016.

Industries Assistance Commission, *Report on Southern Bluefin Tuna* (Australian Government Publishing Service 1984).

International Convention for the Conservation of Atlantic Tunas (Rio de Janeiro, 14 May 1966) 673 UNTS 63.

International Convention for the Regulation of Whaling (Washington, 2 December 1946) 161 UNTS 72.

Kambona, J.J., and S.H. Marashi, *Process for the Establishment of the Indian Ocean Tuna Commission* (FAO Fisheries Circular No 913; FAO 1996).

Kawano, M., 'L'affaire du Thon à nageoire bleue et les chevauchements de juridictions internationales' (2003) XLIX *Annuaire Français de Droit International* 516.

Kaye, S.B., *International Fisheries Management* (Kluwer Law International 2001).

Kennedy, J.O.S., L. Davies and A. Cox, *Joint Rent Maximisation and Open Access Competition in the Southern Bluefin Tuna Fishery* (Australian Bureau of Agricultural and Resource Economics Conference Paper 99.1) <http://data.daff.gov.au/brs/data/warehouse/pe_abarebrs99000382/PR11266.pdf> accessed 6 August 2015.

Kirkwood, G.P., 'Estimation of von Bertalanffy Growth Curve Parameter using both Length Increment and Age-Length Data' (1983) 40 *Canadian Journal of Fisheries and Aquatic Sciences* 1405.

Koers, A.W., *International Regulation of Marine Fisheries: A Study of Regional Fisheries Organizations* (Fishing News (Books) Ltd 1973).

Kwiatkowska, B., 'The Australia and New Zealand v. Japan Southern Bluefin Tuna (Jurisdiction and Admissibility) Award of the First Law of the Sea Convention Annex VII Arbitral Tribunal' (2001) 16 *International Journal of Marine and Coastal Law* 239.

Kwiatkowska, B., 'The Southern Bluefin Tuna Arbitral Tribunal Did Get It Right: A Commentary and Reply to the Article by David A. Colson and Dr. Peggy Hoyle' (2003) 34 *Ocean Development & International Law* 369.

Maguire, J-J., 'Southern Bluefin Tuna Dispute' in M.H. Nordquist and J. Norton Moore (eds), *Current Fisheries Issues and the Food and Agriculture Organization of the United Nations* (Martinus Nijhoff Publishers 2000) 201.

Majkowski, J., H. Arrizabalaga, F. Carocci and H. Murua, 'Tuna and Tuna-like Species' in FAO, *Review of the state of world marine fishery resources* (FAO Fisheries and Aquaculture Technical Paper 569, FAO 2011) 227, http://www.fao.org/docrep/015/i2389e/i2389e.pdf.

Mansfield, B., 'The Southern Bluefin Tuna Arbitration: Comments on Professor Barbara Kwiatkowska's Article' (2001) 16 *International Journal of Marine and Coastal Law* 361.

Mansfield, B., 'Compulsory Dispute Settlement after the Southern Bluefin Tuna Award' in A.G. Oude Elferink and D.R. Rothwell (eds), *Oceans Management in the 21st Century: Institutional Frameworks and Responses* (Martinus Nijhoff Publishers 2004) 255.

Matsuda, Y., 'Postwar Development and Expansion of Japan's Tuna Fishery' in D.J. Doulman (ed), *Tuna Issues and Perspectives in the Pacific Islands Region* (Pacific Islands Development Program 1987) 71.

Ministry of Foreign Affairs of the People's Republic of China, 'Position Paper of the Government of the People's Republic of China on the Matter of Jurisdiction in the South China Sea Arbitration Initiated by the Republic of the Philippines' (7 December 2014) <http://www.fmprc.gov.cn/mfa_eng/zxxx_662805/t1217147.shtml> accessed 2 April 2016.

Morgan, D.L., 'A Practitioner's Critique of the Order Granting Provisional Measures in the Southern Bluefin Tuna Cases' in M.H. Nordquist and J. Norton Moore (eds), *Current Marine Environmental Issues and the International Tribunal for the Law of the Sea* (Martinus Nijhoff Publishers 2001) 173.

Morgan, D.L., 'Implications of the Proliferation of International Legal Fora: The Example of the Southern Bluefin Tuna Cases' (2002) 43 *Harvard International Law Journal* 541.

MOX Plant Case (Ireland v. United Kingdom) (Provisional Measures) [2001] ITLOS Rep 95.

Munro, G.R., and A.D. Scott, 'The Economics of Fisheries Management' in A.V. Kneese and J.L. Sweeney (eds), *Handbook of Natural Resource and Energy Economics*, vol II (Elsevier 1985) 623.

Nordquist, M.H. (Editor-in-Chief), *United Nations Convention on the Law of the Sea, 1982: A Commentary*, vol II (S.N. Nandan, S. Rosenne (vol eds) and N.R. Grandy (assistant ed), Martinus Nijhoff Publishers 1993); vol III (S.N. Nandan and S. Rosenne (vol eds), Martinus Nijhoff Publishers 1995); vol v (S. Rosenne and L.B. Sohn (vol eds), Martinus Nijhoff Publishers 1989).

Orrego Vicuña, F., 'From the 1893 *Bering Sea Fur Seals Case* to the 1999 *Southern Bluefin Tuna Cases*: A Century of Efforts at Conservation of the Living Resources of the High Seas' (1999) 10 *Yearbook of International Environmental Law* 40.

Orrego Vicuña, F., *The Changing International Law of High Seas Fisheries* (Cambridge University Press 1999).

Otani, Y., 'Quelques réflexions sur la juridiction et la recevabilité vis-à-vis de l'*Affaire du thon à nageoire bleue*' in N. Ando, E. McWhinney and R. Wolfrum (eds), *Liber Amicorum Judge Shigeru Oda* (Kluwer Law International 2002) 731.

Oxman, B.H., 'Complementary Agreements and Compulsory Jurisdiction' (2001) 95 *American Journal of International Law* 277.

Peel, J., 'A Paper Umbrella Which Dissolves in the Rain? The Future for Resolving Fisheries Disputes under UNCLOS in the Aftermath of the Southern Bluefin Tuna Arbitration' (2002) 3 *Melbourne Journal of International Law* 53.

Pitcher, T.J., and P.J.B. Hart, *Fisheries Ecology* (Croom Helm 1982).

Polacheck, T., 'An overview of interaction issues among the fisheries for Southern Bluefin Tuna' in R.S. Shomura, J. Majkowski and S. Langi (eds), *Interactions of Pacific Tuna Fisheries: Proceedings of the First FAO Expert Consultation on Interactions of Pacific Tuna Fisheries* (3–11 December 1991) vol 1 (FAO Technical Paper 336/1, FAO 1994) 264.

Polacheck, T., 'Experimental catches and the precautionary approach: the Southern Bluefin Tuna dispute' (2002) 26 *Marine Policy* 283.

Polacheck, T., N.L. Klaer, C. Millar and A.L. Preece, 'An initial variation of management strategies for the southern bluefin tuna fishery' (1999) 56 *ICES* [*International Council for the Exploration of the Sea*] *Journal of Marine Science* 811.

Pulp Mills on the River Uruguay (Argentina v. Uruguay) Judgment [2010] ICJ Rep 14.

Ricker, W.E., 'Stock and Recruitment' (1954) 11 *Journal of the Fisheries Research Board of Canada* 559.

Röben, V., 'The Southern Bluefin Tuna Cases: Re-Regionalization of the Settlement of Law of the Sea Disputes?' (2002) 62 *Zeitschrift für ausländisches öffentliches Recht und Völkerrecht* 61.

Romano, C., 'The *Southern Bluefin Tuna* Dispute: Hints of a World to Come … Like It or Not' (2001) 32 *Ocean Development & International Law* 313.

Rothschild, B.J., and A. Suda, 'Population Dynamics of Tuna' in J. Gulland (ed), *Fish Population Dynamics* (John Wiley & Sons 1977) 309.

Schaefer, M.B., 'Some Considerations of Population Dynamics and Economics in Relation to the Management of Commercial Fisheries' (1957) 14 *Journal of the Fisheries Research Board of Canada* 669.

Scheiber, H.N., 'Origins of the Abstention Doctrine in Ocean Law: Japanese-U.S. Relations and the Pacific Fisheries, 1937–1958' (1989) 16 *Ecology Law Quarterly* 23.

Serdy, A., 'One fin, two fins, red fins, bluefins: some problems of nomenclature and taxonomy affecting legal instruments governing tuna and other highly migratory species' (2004) 28 *Marine Policy* 235.

Serdy, A., 'The Paradoxical Success of UNCLOS Part XV: A Half-Hearted Reply to Rosemary Rayfuse' (2005) 36 *Victoria University of Wellington Law Review* 713.

Serdy, A., 'Accounting for Catch in Internationally Managed Fisheries: What Role for State Responsibility?' (2010) 15 *Ocean and Coastal Law Journal* 23.

Serdy, A., 'Implementing Article 28 of the UN Fish Stocks Agreement: The First Review of a Conservation Measure in the South Pacific Regional Fisheries Management Organisation' (2016) 47 *Ocean Development & International Law* 1.

Serdy, A., *The New Entrants Problem in International Fisheries Law* (Cambridge University Press 2016).

Southern Bluefin Tuna Case (Australia and New Zealand v. Japan) (Award on Jurisdiction and Admissibility, 4 August 2000) XXIII RIAA 1.

Southern Bluefin Tuna Case (Australia and New Zealand v. Japan) Government of Japan, *Cases Concerning the Convention for the Conservation of Southern Bluefin Tuna* [*sic*] (Memorial on Jurisdiction of Japan) <https://icsid.worldbank.org/apps/ICSIDWEB/Documents/Memorial%20on%20Jurisdiction%20of%20Japan.pdf> accessed 5 August 2015.

Southern Bluefin Tuna Case (Australia and New Zealand v. Japan) Reply on Jurisdiction [of] Australia and New Zealand <https://icsid.worldbank.org/apps/ICSIDWEB/Documents/Reply%20on%20Jurisdiction%20of%20Australia%20and%20New%20Zealand.pdf> accessed 6 August 2015.

Southern Bluefin Tuna Cases (New Zealand v Japan; Australia v Japan) (Provisional Measures, Order of 27 August 1999) [1999] ITLOS Rep 280.

Southern Bluefin Tuna Cases (New Zealand v Japan; Australia v Japan) Statements of Claim under art 1 of Annex VII to UNCLOS <https://icsid.worldbank.org/apps/ICSIDWEB/Documents/Statement%20of%20Claim%20of%20Australia%20and%20New%20Zealand.pdf> accessed 6 August 2015.

Southern Bluefin Tuna Cases (New Zealand v Japan; Australia v Japan) Response and Counter-Request for Provisional Measures submitted by Japan, ITLOS doc F4/155/198 <https://www.itlos.org/fileadmin/itlos/documents/cases/case_no_3_4/statement_response_japan_eng.pdf> accessed 14 August 2015.

Stephens, T., 'The Limits of International Adjudication in International Environmental Law: Another Perspective on the Southern Bluefin Tuna Case' (2004) 19 *International Journal of Marine and Coastal Law* 177.

Tanaka, N., 'Some Observations on the Southern Bluefin Tuna Arbitration Award' (2001) 44 *Japanese Annual of International Law* 9.

Thorogood, J., 'Age and Growth Rate Determination of Southern Bluefin Tuna, *Thunnus maccoyii*, Using Otolith Banding' (2006) 30 *Journal of Fish Biology* 7.

Treaty of Peace with Japan (San Francisco, 8 September 1951) 136 UNTS 45.

United Nations Convention on the Law of the Sea (Montego Bay, 10 December 1982) 1833 UNTS 3.

Whaling in the Antarctic (*Australia v. Japan: New Zealand intervening*) (Judgment) [2014] ICJ Rep 226.

Williams, S., 'Understanding Japanese seafood markets' (1992) 51(2) *Australian Fisheries* 32.

www.ingramcontent.com/pod-product-compliance
Lightning Source LLC
Chambersburg PA
CBHW031331230926
51848CB00001B/181
9789004339446